AGRICULTURAL RESOURCE USE AND MANAGEMENT

AGRICULTURAL RESOURCE USE AND MANAGEMENT

Edited by

Kimberly Etingoff

Apple Academic Press

TORONTO NEW JERSEY

Apple Academic Press Inc.
3333 Mistwell Crescent
Oakville, ON L6L 0A2
Canada

Apple Academic Press Inc.
9 Spinnaker Way
Waretown, NJ 08758
USA

Exclusive worldwide distribution by CRC Press, a member of Taylor & Francis Group

No claim to original U.S. Government works
Printed in the United States of America on acid-free paper

International Standard Book Number-13: 978-1-77188-073-2 (Hardcover)

Library of Congress Control Number: 2014943672

Library and Archives Canada Cataloguing in Publication

Agricultural resource use and management/edited by Kimberly Etingoff.

Includes bibliographical references and index.
ISBN 978-1-77188-073-2 (bound)
1. Agriculture--Environmental aspects. 2. Agricultural innovations. I. Etingoff, Kim, editor

S589.75.A37 2014   630.28'6   C2014-904489-5

Apple Academic Press also publishes its books in a variety of electronic formats. Some content that appears in print may not be available in electronic format. For information about Apple Academic Press products, visit our website at **www.appleacademicpress.com** and the CRC Press website at **www.crcpress.com**

ABOUT THE EDITOR

KIMBERLY ETINGOFF

Kimberly Etingoff's background includes city and regional planning, farming, food systems programming, sociology, and urban geography. She studied at the University of Rochester and Tufts University and has done extensive field work with food systems and agricultural sociology. She has been writing and editing academic and educational books on topics such as nutrition, farming, and aspects of sociology for more than ten years.

CONTENTS

ACKNOWLEDGMENT AND HOW TO CITE

The editor and publisher thank each of the authors who contributed to this book, whether by granting their permission individually or by releasing their research as open source articles or under a license that permits free use, provided that attribution is made. The chapters in this book were previously published in various places in various formats. To cite the work contained in this book and to view the individual permissions, please refer to the citation at the beginning of each chapter. Each chapter was read individually and carefully selected by the editor; the result is a book that provides a nuanced study of the agricultural resource management. The chapters included examine the following topics:

- Chapter 1 establishes a connection between chemical fertilizer use in agriculture and stream water quality on several scales.
- The authors of Chapter 2 present an overview and evaluation of three technologies that can be implemented to reuse wastewater resources, along with nutrients and energy, for agriculture in peri-urban areas in particular
- Chapter 3 analyzes several options for reducing the loss of soil nitrogen from cole crop production in the form of carbon amendments, increasing agricultural land viability.
- Chapter 4 describes how both organic and non-organic cattle grazing systems have different effects on pasture quality, which in turn affects cattle health and development.
- Chapter 5 argues that long-term agricultural activities negatively impact bacterial diversity in soil, an important indicator in overall ecological health, when compared with bacterial diversity in adjacent non-agricultural land
- Chapter 6 introduces a model for estimating the carbon footprint of livestock production, which can be used to connect land use management practice to carbon reduction strategies.
- The authors of Chapter 7pursue the claim that there exist important opportunities to decrease diesel intensity and improve energy consumption rates in agriculture.
- Precision agriculture, described in Chapter 8, represents one method of reducing resource inputs in agriculture, specifically energy-based inputs such as irrigation and fertilizers, and making resource use more efficient.

- Chapter 9 investigates irrigation management techniques that can improve not only the depletion of water resources, but also energy use and greenhouse gas emissions.
- Chapter 10 argues that people around the world will need to adapt to the pressures a growing population will have on agricultural systems and resources.
- The authors of Chapter 11 argue that priority should be given to crop breeding programs that focus on improving production yields of low-input agriculture, as a response to the growing agricultural energy and resource demands created by a larger global population.
- A combination of an increasing population and a slowing of the growth of agricultural outputs has resulted in stress on the West Bengal agricultural sector; Chapter 12 explores the specific dynamics of these trends.

LIST OF CONTRIBUTORS

Gary L. Andersen
Department of Ecology, Lawrence Berkeley National Laboratory, Berkeley, California, United States of America

Iemke Bisschops
Lettinga Associates Foundation, Bornse weilanden 9, 6700 AM Wageningen, The Netherlands

Vagner A. Benedito
Division of Plant & Soil Sciences, P.O. Box 6108, West Virginia University, Morgantown, WV 26506, USA

Ganesh C. Bora
Department of Agricultural and Biosystems Engineering, North Dakota State University, 1221 Albrecht Blvd, Fargo, ND 58102, USA

Angel Carrillo
Centro de Investigaciones biologicas del Noroeste, S.C. La Paz, Mexico

Thelma Castellanos
Centro de Investigaciones biologicas del Noroeste, S.C. La Paz, Mexico

Katelyn A. Congreves
School of Environmental Sciences, University of Guelph, Ridgetown Campus, Ridgetown, Ontario, N0P 2C0, Canada

Raymond L. Desjardins
Eastern Cereal and Oilseed Research Centre, Agriculture and Agri-Food Canada (AAFC), Ottawa, ON, K1A 0C6, Canada

Guo-Chun Ding
Institute for Epidemiology and Pathogen Diagnostics - Federal Research Centre for Cultivated Plants (JKI), Braunschweig, Germany

Anja B. Dohrmann
Institute for Biodiversity, Johann Heinrich von Thünen-Institut (TI), Braunschweig, Germany

James A. Dyer
AAFC Consultant, Cambridge, ON, N3H 3Z9, Canada

Laura Essl
Centre for Environmental Management and Decision Support, Vienna 1180, Austria

Tiffany L. Fess
Division of Plant & Soil Sciences, P.O. Box 6108, West Virginia University, Morgantown, WV 26506, USA

Shusen Gui
Key Laboratory of Ocean Energy Utilization and Energy Conservation of Ministry of Education, Dalian University of Technology, Dalian 116024, China

Aimee N. Hafla
United States Department of Agriculture, Agriculture Research Service, Pasture Systems and Watershed Management Research Unit, Building 3702, Curtin Road, University Park, PA 16802, USA

Holger Heuer
Institute for Epidemiology and Pathogen Diagnostics - Federal Research Centre for Cultivated Plants (JKI), Braunschweig, Germany

A. H. M. Zehadul Karim
Department of Sociology and Anthropology, International Islamic University Malaysia, Gombak, Kuala Lumpur, Malaysia

James B. Kotcon
Division of Plant & Soil Sciences, P.O. Box 6108, West Virginia University, Morgantown, WV 26506, USA

Hong Li
Institute of Agricultural Integrated Development, Beijing Academy of Agriculture and Forestry Sciences, No.9 Shu Guang Hua Yuan Middle Road, Beijing 100097, China

Huanan Li
Key Laboratory of Ocean Energy Utilization and Energy Conservation of Ministry of Education, Dalian University of Technology, Dalian 116024, China

Nan Li
Key Laboratory of Ocean Energy Utilization and Energy Conservation of Ministry of Education, Dalian University of Technology, Dalian 116024, China

Eduardo López
The Mexican Institute of Water Technology (IMTA), Cuernavaca 62550, México

Jennifer W. MacAdam
Utah State University, Department of Plants, Soils and Climate, 4820 Old Main Hill, Logan 84322, UT, USA

José Luis Martínez-Ruiz
The Mexican Institute of Water Technology (IMTA), Cuernavaca 62550, México

Brian G. McConkey
Semiarid Prairie Agricultural Research Centre, Agriculture and Agri-Food Canada (AAFC), Swift Current, SA, S9H 3X2, Canada

Kasturi Mondal
International Institute for Population Sciences, Mumbai, India

Hailin Mu
Key Laboratory of Ocean Energy Utilization and Energy Conservation of Ministry of Education, Dalian University of Technology, Dalian 116024, China

Daniel Murillo
The Mexican Institute of Water Technology (IMTA), Cuernavaca 62550, México

Tiemen A. Nanninga
Lettinga Associates Foundation, Bornse weilanden 9, 6700 AM Wageningen, The Netherlands

John F. Nowatzki
Department of Agricultural and Biosystems Engineering, North Dakota State University, 1221 Albrecht Blvd, Fargo, ND 58102, USA

Yvette M. Piceno
Department of Ecology, Lawrence Berkeley National Laboratory, Berkeley, California, United States of America

Asad Sarwar Qureshi
Senior Environment Specialist, National Development Consultants (NDC), Lahore, Pakistan

David C. Roberts
Department of Agribusiness and Applied Economics, North Dakota State University, 500 Barry Hall, Fargo, ND 58102, USA

Sanjit Sarkar
International Institute for Population Sciences, Mumbai, India

Kornelia Smalla
Institute for Epidemiology and Pathogen Diagnostics - Federal Research Centre for Cultivated Plants (JKI), Braunschweig, Germany

Ward N. Smith
Eastern Cereal and Oilseed Research Centre, Agriculture and Agri-Food Canada (AAFC), Ottawa, ON, K1A 0C6, Canada

Kathy J. Soder
United States Department of Agriculture †, Agriculture Research Service, Pasture Systems and Watershed Management Research Unit, Building 3702, Curtin Road, University Park, PA 16802, USA

Markus Starkl
Competence Centre for Decision-Aid in Environmental Management, University of Natural Resources and Life Sciences/DIB, Gregor Mendel Strasse 33, Vienna 1180, Austria

Danfeng Sun
College of Natural Resources and Environmental Science, China Agricultural University, Beijing 100094, China

Christoph C. Tebbe
Institute for Biodiversity, Johann Heinrich von Thünen-Institut (TI), Braunschweig, Germany

Laura L. Van Eerd
School of Environmental Sciences, University of Guelph, Ridgetown Campus, Ridgetown, Ontario, N0P 2C0, Canada

Xavier P.C. Vergé
AAFC Consultant, Ottawa, ON, K2A 1G6, Canada

Richard J. Vyn
Department of Food, Agriculture and Resource Economics, University of Guelph, Ridgetown Campus, Ridgetown, Ontario, N0P 2C0, Canada

Nicole Weinert
Institute for Epidemiology and Pathogen Diagnostics - Federal Research Centre for Cultivated Plants (JKI), Braunschweig, Germany

Devon E. Worth
Eastern Cereal and Oilseed Research Centre, Agriculture and Agri-Food Canada (AAFC), Ottawa, ON, K1A 0C6, Canada

Weiwei Zhang
Institute of Agricultural Integrated Development, Beijing Academy of Agriculture and Forestry Sciences, No.9 Shu Guang Hua Yuan Middle Road, Beijing 100097, China

Liandi Zhou
Institute of Agricultural Integrated Development, Beijing Academy of Agriculture and Forestry Sciences, No.9 Shu Guang Hua Yuan Middle Road, Beijing 100097, China

INTRODUCTION

Agriculture is one of the world's greatest sources of resource consumption. Globally, modern agricultural activities use vast amounts of water, soil nutrients, and, particularly, energy inputs. The effects of resource use are far-reaching, from greenhouse gas emissions to localized ecosystem imbalances. Agricultural resource use issues are of particular concern in today's world, as we are faced with climate change and a growing population. Many researchers recognize this and have undertaken the task of analyzing agricultural resource use patterns and particularities. In addition, scientists have developed and studied resource management methods that have the potential to address resource use and mitigate the detrimental effects resource use has on people and the natural environment.

Kimberly Etingoff

Understanding the effects of intensive agricultural land use activities on water resources is essential for natural resource management and environmental improvement. In Chapter 1, by Zhang and colleagues, multi-scale nested watersheds were delineated and the relationships between two representative water quality indexes and agricultural land use intensity were assessed and quantified for the year 2000 using multi-scale regression analysis. The results show that the log-transformed nitrate-nitrogen (NO_3-N) index exhibited a relationship with chemical fertilizer input intensity and several natural factors, including soil loss, rainfall and sunlight at the first order watershed scale, while permanganate index (CODMn) had a positive relationship with another two input intensities of pesticides and agricultural plastic mulch and organic manure at the fifth order watershed scale. The first order watershed and the fifth order watershed were considered as the watershed adaptive response units for NO_3-N and CODMn, respectively. The adjustment of agricultural input and its intensity may be carried out inside the individual watershed adaptive response unit. The multiple linear regression model demonstrated the cause-and-effect relationship between agricultural land use intensity and stream water quality at multiple scales, which is an important factor for the maintenance of stream water quality.

Often, centralized water supply, sanitation and solid waste services struggle to keep up with the rapid expansion of urban areas. The peri-urban areas are at the forefront of this expansion and it is here where decentralized technologies are increasingly being implemented. The introduction of decentralized technologies allows for the development of new opportunities that enable the recovery and reuse of resources in the form of water, nutrients and energy. This resource-oriented management of water, nutrients and energy requires a sustainable system aimed at low resource use and high recovery and reuse rates. Instead of investigating each sector separately, as has been traditionally done, in Chapter 2 Nanninga and colleagues propose and discuss a concept that seeks to combine the in- and outflows of the different sectors, reusing water and other liberated resources where possible. This paper shows and demonstrates examples of different types of sustainable technologies that can be implemented in the biofiltros (small constructed wetlands), and (vermi-)composting]. An innovative participatory planning method, combining scenario development with a participatory planning workshop with key stakeholders, was applied and resulted in three concept scenarios. Specific technologies were then selected for each concept scenario that the technical feasibility and applicability was assessed. Following this, the resulting resource flows (nutrients, water and energy) were determined and analyzed. The results show that decentralized technologies not only have the potential to deliver adequate water supply, sanitation and solid waste services in peri-urban areas and lessen environmental pollution, but also can recover significant amounts of resources thereby saving costs and providing valuable inputs in, for instance, the agricultural sector. Social acceptance of the technologies and institutional cooperation, however, is key for successful implementation.

Cole crops (*Brassica* vegetables) can pose a significant risk for N losses during the post-harvest period due to substantial amounts of readily mineralizable N in crop residues. Amending the soil with organic C has the potential to immobilize N and thereby reduce the risk for N losses. In Chapter 3, Congreves and colleagues conducted four field trials to determine the effects of organic C amendments (OCA) on N dynamics and spring wheat (*Triticum durum* L.) harvest parameters proceeding early- and late-broccoli (*Brassica olecerea* var *italica* L.) systems in 2009 and 2010. The experimental controls represented the traditional grower practice of incorporated broccoli crop residue (CR-control) and the pre-plant application of N fertilizer (CRN-control) to subsequent spring wheat. Alternative practices were compared to the controls, which included broccoli crop residue removal (CR-removal), an oat (*Avena sativa* L.) cover crop (CC-oat), and three different OCA of wheat straw (OCA-straw), yard waste (OCA-yard), or used cooking oil (OCA-oil). The treatments, which demonstrated reduced autumn soil mineral N (SMN) concentrations after broccoli harvest, relative to the CR-control, were CR-removal, OCA-straw, and OCA-oil. Although CR-removal and OCA-straw indicated a reduced potential for autumn soil N losses in the early-broccoli system, these practices are not recommended for growers because subsequent spring wheat yield and profit margins were reduced compared to the CR- and CRN-controls. The OCA-oil reduced autumn SMN concentrations by 53 to 112 kg N ha^{-1} relative to the CR-control after both early- and late-broccoli harvest, suggesting a larger potential for reduced autumn soil N losses, compared to all other treatments. No detrimental effects resulted from the OCA-oil treatment on the subsequent spring yield or grain N. The OCA-oil reduced spring wheat profit margins relative to the CR-control, like the OCA-straw and CR-removal treatments, however profit margins were similar between the OCA-oil and the CRN-control. Therefore, in areas with a high risk of environmental N contamination, growers should consider the OCA-oil practice after cole crop harvest to minimize the risk of N losses.

In 2010, the National Organic Program implemented a rule for the US stating that pasture must be a significant source of feed in organic ruminant systems. Chapter 4, by Hafla and colleagues focuses on how the pasture rule has impacted the management, economics and nutritional value of products derived from organic ruminant systems and the interactions of grazing cattle with pasture forages and soils. The use of synthetic fertilizers is prohibited in organic systems; therefore, producers must rely on animal manures, compost and cover crops to increase and maintain soil nitrogen content. Rotational and strip grazing are two of the most common grazing management practices utilized in grazing ruminant production systems; however, these practices are not exclusive to organic livestock producers. For dairy cattle, grazing reduces foot and leg problems common in confinement systems, but lowers milk production and exposes cows to parasites that can be difficult to treat without pharmaceuticals. Organic beef cattle may still be finished in feedlots for no more than 120 days in the US, but without growth hormones and antibiotics, gains may be reduced and illnesses increased. Grazing reduces the use of environmentally and economically costly concentrate feeds and recycles nutrients back to the soil efficiently, but lowers the rate of beef liveweight gain. Increased use of pasture can be economically, environmentally and socially sustainable if forage use efficiency is high and US consumers continue to pay a premium for organic beef and dairy products.

Natural scrublands in semi-arid deserts are increasingly being converted into fields. This results in losses of characteristic flora and fauna, and may also affect microbial diversity. In Chapter 5, Ding and colleagues explored the long-term effect (50 years) of such a transition on soil bacterial communities at two sites typical of semi-arid deserts. Comparisons were made between soil samples from alfalfa fields and the adjacent scrublands by two complementary methods based on 16S rRNA gene fragments amplified from total community DNA. Denaturing gradient gel electrophoresis (DGGE) analyses revealed significant effects of the transition on community composition of *Bacteria, Actinobacteria, Alpha-* and *Betaproteobacteria* at both sites. PhyloChip hybridization analysis uncovered that the transition negatively affected taxa such as *Acidobacteria, Chloroflexi, Acidimicrobiales, Rubrobacterales, Deltaproteobacteria* and *Clostridia,* while *Alpha-, Beta-* and *Gammaproteobacteria, Bacteroidetes* and *Actinobacteria* increased in abundance. Redundancy analysis suggested that the community composition of phyla responding to agricultural use (except for *Spirochaetes*) correlated with soil parameters that were significantly different between the agricultural and scrubland soil. The arable soils were lower in organic matter and phosphate concentration, and higher in salinity. The variation in the bacterial community composition was higher in soils from scrubland than from agriculture, as revealed by DGGE and

PhyloChip analyses, suggesting reduced beta diversity due to agricultural practices. The long-term use for agriculture resulted in profound changes in the bacterial community and physicochemical characteristics of former scrublands, which may irreversibly affect the natural soil ecosystem.

To assess tradeoffs between environmental sustainability and changes in food production on agricultural land in Canada the Unified Livestock Industry and Crop Emissions Estimation System (ULICEES) was developed. It incorporates four livestock specific GHG assessments in a single model. To demonstrate the application of ULICEES, 10% of beef cattle protein production was assumed to be displaced with an equivalent amount of pork protein. Without accounting for the loss of soil carbon, this 10% shift reduced GHG emissions by 2.5 $TgCO_2e$ y^{1}. The payback period was defined as the number of years required for a GHG reduction to equal soil carbon lost from the associated land use shift. A payback period that is shorter than 40 years represents a net long term decrease in GHG emissions. Displacing beef cattle with hogs resulted in a surplus area of forage. When this residual land was left in ungrazed perennial forage, the payback periods were less than 4 years and when it was reseeded to annual crops, they were equal to or less than 40 years. They were generally greater than 40 years when this land was used to raise cattle. Agricultural GHG mitigation policies will inevitably involve a trade-off between production, land use and GHG emission reduction. Chapter 6, by Vergé and colleagues, argues that ULICEES is a model that can objectively assess these trade-offs for Canadian agriculture.

As agricultural mechanization accelerates the development of agriculture in China, to control the growth of the resulting energy consumption of mechanized agriculture without negatively affecting economic development has become a major challenge. A systematic analysis of the factors (total power, unit diesel consumption, etc.) influencing diesel consumption using the SECA model, combined with simulations on agricultural diesel flows in China between 1996 and 2010 is performed in Chapter 7, by Li and colleagues. Seven agricultural subsectors, fifteen categories of agricultural machinery and five farm operations are considered. The results show that farming and transportation are the two largest diesel consumers, accounting for 86.23% of the total diesel consumption in agriculture in 2010. Technological progress has led to a decrease in the unit diesel consumption and an increase in the unit productivity of all machinery, and there is still much potential for future progress. Additionally, the annual average working hours have decreased rapidly for most agricultural machinery, thereby influencing the development of mechanized agriculture.

Energy input in agriculture has increased tremendously and accounts for about 17% of total energy consumed in the USA. Precision agriculture involves knowledge-based technical management systems to optimize application of fertilizer, chemicals, seeds, and irrigation resources to reduce input costs and to enhance crop yield while simultaneously reducing harmful environmental impacts associated with inefficient use of agricultural inputs. It also uses GPS-based auto-guidance systems in agricultural vehicles to reduce overlapping of equipment and tractor passes, thus saving fuel, labor, time, and soil compaction with environmental benefit. The study in Chapter 8 was undertaken by Bora and colleagues to quantify the fuel and labor savings resulting from adoption of precision agricultural technology in the upper mid-west state of North Dakota in the USA. A survey was conducted with responses from farmers of various demographics about savings of time and fuel in their agricultural vehicle by the use of GPS guidance and autosteering systems. It was found that 34% farms used GPS guidance systems, reducing machine time and fuel consumption by 6.04% and 6.32%, respectively. Twenty-seven percent of the farms used autosteering systems, which further reduced machine time by 5.75% and fuel consumption by 5.33%. GPS guidance and autosteering systems can save an average of 1,647 and 1,866 L of fuel per farm respectively. The monetized values of time saved for the average farm are US733.85 and US851.27 for GPS guidance and autosteering systems, respectively. The farm energy savings in terms of fuel and time by using GPS guidance and/or autosteering systems in farm vehicles in the Upper-Midwest region of the USA was estimated from the survey results. Based on the perceptions of farmers adopting precision agriculture, the two technologies investigated in this research provided a positive return on investment and would be beneficial to North Dakota's agricultural sector if adopted more widely.

Increasing use of groundwater for irrigation is linked to high energy demand, depleting resources and resulting in a high carbon footprint. Chapter 9, by Qureshi, explores how improved on-farm irrigation management can help in reducing groundwater extraction, limiting energy consumption and CO_2 emissions. In Pakistan, every year about 50 billion cubic metres (BCM) of groundwater is pumped for irrigation, which consumes more than 6 billion kWh of electricity and 3.5 billion litres of diesel. Carbon emissions attributed to this energy use amount to 3.8 million metric tons (MMT) of CO_2 per year. Considerable research carried out in Pakistan has suggested that improved irrigation management can significantly reduce the irrigation water applied to different crops. This study revealed that by adopting improved irrigation schedules, water productivity will increase and groundwater withdrawals for irrigation can be reduced by 24 BCM. Reduced groundwater extraction will result in a 62% decline in energy demand (1.5 billion litres of diesel as most of the private tubewells run on diesel) and a 40% reduction in carbon emissions. In addition, a reduction in irrigation applications will also be beneficial for stabilizing groundwater tables and groundwater quality.

A decade ago, David Pimentel and his associates (1998) reported to Karim, the author of Chapter 10, that at least ten million hectares of arable land were being eroded and also abandoned throughout the world every year and consequently to compensate such a loss, a huge amount of replacement is claimed from forests and other sources for agriculture and human settlement. In the meantime, world population exceeded 6 billion in the year 1999, and the projected data indicate that it is going to be almost 9 billion within the next 40 years. For that reason, the demographers and environmentalists have highlighted that the main challenge for environmental management throughout the world today is to determine our planet's capacity to sustain such a huge amount of burgeoning human population. The paper thus assesses specifically the impact of growing population on agricultural resources around the world, creating depressing pressure on sustainable environmental management. To exemplify such a trend of agricultural land use, the paper incorporates a detailed example from an ethnographic case study on indigenous land-use practices and the experiences associated with modern cultivation for adapting to adverse situations caused by severe impact of a growing population in the agricultural sector in rural Bangladesh.

World population is projected to reach its maximum (~10 billion people) by the year 2050. This 45% increase of the current world population (approaching seven billion people) will boost the demand for food and raw materials. However, we live in a historical moment when supply of phosphate, water, and oil are at their peaks. Modern agriculture is fundamentally based on varieties bred for high performance under high input systems (fertilizers, water, oil, pesticides), which generally do not perform well under low-input situations. In Chapter 11, Fess and colleagues propose a shift of research goals and plant breeding objectives from high-performance agriculture at high-energy input to those with an improved rationalization between yield and energy input. Crop breeding programs that are more focused on nutrient economy and local environmental fitness will help reduce energy demands for crop production while still providing adequate amounts of high quality food as global resources decline and population is projected to increase.

Chapter 12, by Sarkar and Mondal, examines the dynamics in population growth and sustainability of food grain production in West Bengal. Linkage between population growth and food production is an issue of debate since late eighteenth century when Malthus predicted that population growth will outstrip the food supply. Though fertility level in West Bengal reached to below replacement (TFR is <2.1) but population will increase till next few decades due to the mechanism of population momentum. Average annual growth rate has declined over the last two decades but absolute growth in the population increases the demand for food. There has been remarkable increase in the food grain production in West Bengal after 1980s but till the current level of food production is not sufficient enough to meet the domestic food requirement, though this gap decreased over the time. Besides, slow growth in the agricultural in the last few years is another concern of sustainable food production. Population growth in the West Bengal has significant association with food grain production and agriculture. Cultivable land and net sown area has reduced significantly due to the rapid growth of population. Cropping intensity increased drastically because of the reduction of net sown area and increase in population. Hence, it is very essential to increase the current level of food production more than proportional of population growth to ensure the food security in the near future in West Bengal.

PART I

WATER

CHAPTER 1

A STATISTICAL ASSESSMENT OF THE IMPACT OF AGRICULTURAL LAND USE INTENSITY ON REGIONAL SURFACE WATER QUALITY AT MULTIPLE SCALES

WEIWEI ZHANG, HONG LI, DANFENG SUN, AND LIANDI ZHOU

1.1 INTRODUCTION

Land use intensity is one of the most significant forms of land cover modification, and can have a major detrimental impact on terrestrial and aquatic ecosystems [1,2], and also directly influence human and ecosystem health. Many developed countries are experiencing environmental pollution due to intensive agricultural activities, including intensive crop and livestock production [3]. This is also the true for the fast developing countries, such as China. From a land use perspective, intensive agricultural activities have been identified as the major sources of non-point source pollutants and are known to alter and impact the quality of the receiving water bodies. As an environmental factors that relate directly to human health, water quality is always subject to degradation when agricultural land use intensity is too

---

This chapter was originally published under the Creative Commons Attribution License. Zhang W, Li H, Sun D, and Zhou L. A Statistical Assessment of the Impact of Agricultural Land Use Intensity on Regional Surface Water Quality at Multiple Scales. International Journal of Environmental Research and Public Health ***9*** *(2012); doi:10.3390/ijerph9114170.*

high [4]. Thus, understanding the effects of intensive agricultural land use activities on water resources is essential for natural resource management and environmental improvement. However, these effects on water quality conditions are difficult to determine because of the complex relationships between agricultural land use activities and water quality.

In previous studies water quality was generally linked to land use inside the catchments area. Several studies have found that land use has a strong influence on the receiving water body quality [5–7]. The majority of studies about the effects of land use on water quality have focused on either deterministic modeling, or spatial, or statistical analyses. Examples of modeling studies include those performed by Tong and Chen [5], Chaplot et al. [8], Cao et al. [9], Bhattarai et al. [10], etc. which have adopted the existing watershed-scale hydrological variables and nonpoint-source pollution models to evaluate or predict how land use/land cover scenarios affect water quality. Since modeling methods need long-term water quality monitoring data and regional parameters are difficult to obtain, current modeling methods are still developmental and confined to mechanism studies in local watersheds. Consequently, there are more studies that have adopted statistical methods such as correlation analysis [11,12], single linear regression analysis [13,14], multiple linear regression analysis [15–17], nonparametric statistical analysis techniques [18], etc. to examine the relationships between watershed land use/land cover and water quality.

Since no statistical significant relationships between land uses and nitrate level were found when using the whole basins, contributing areas inside buffer zones were developed by Basnyat et al [19]. There have been more subsequent studies taking buffer zones as analysis units to explore water quality characteristics and their relationships [20–22]. The definition of contributing zone may open additional ways of visualizing the problem. The previous studies have demonstrated that the contributing zone is influenced by many factors, including the water-quality parameter being assessed and geomorphic/climatic setting of the watershed [19]. To some extent, buffer zones with multi-scale characteristics, created using the distance from the stream, are not true hydrological units, and they are difficult to delineate and explain the hydrological and ecological condition of the stream validly. To overcome this, our study defines the multi-scale nested watersheds based on the basic watershed units created by a digital elevation model for the purpose of more effective watershed management, and multi-scale analysis is adopted to explore the relationships between agricultural land use intensity and water quality, and further to identify watershed adaptive response units for every water quality parameter.

Beijing's mountainous watersheds, providing 69.9% of its surface water resources, have played increasingly important roles in drinking water supply and headwater conservation considering the population increase and urban sprawl of Beijing. Moreover, land use changes in the Beijing mountainous areas have brought about many land related problems, such as water pollution, soil contamination and air pollution [23]. We had adopted emergy analysis with principal component analysis, regression analysis and cluster classification to investigate the characteristics and patterns of agricultural land use intensity of study areas in 2000, as the baseline of ecological monitoring and assessment [24]. However, the effects of the agricultural land use intensity on surface water quality have not been discussed. Therefore, the objective of this study, taking the Beijing mountainous area as a case, was to investigate the impacts of agricultural land use intensity on selected physical properties of surface water quality using multi-scale analysis for building a baseline database applicable to long-term monitoring.

1.2 MATERIALS AND METHODS

1.2.1 STUDY AREAS

Beijing's mountainous areas, with an area of 1.04×10^6 ha, are located to the west, north and northeast of Beijing. The study areas comprise a total of five rivers, including the Yongding River, Chaobai River, Beiyun River, Jiyun River and Daqing River (Figure 1). Mean annual precipitation in the area is about 566 mm, about 60% of which falls in July and August. The annual average evaporation is about 1,761 mm. Annual average runoff was about 1.8×10^9 m^3, but this had decreased to 1.3×10^9 m^3 by the end of the last century as a result of climate and land use/land cover changes.

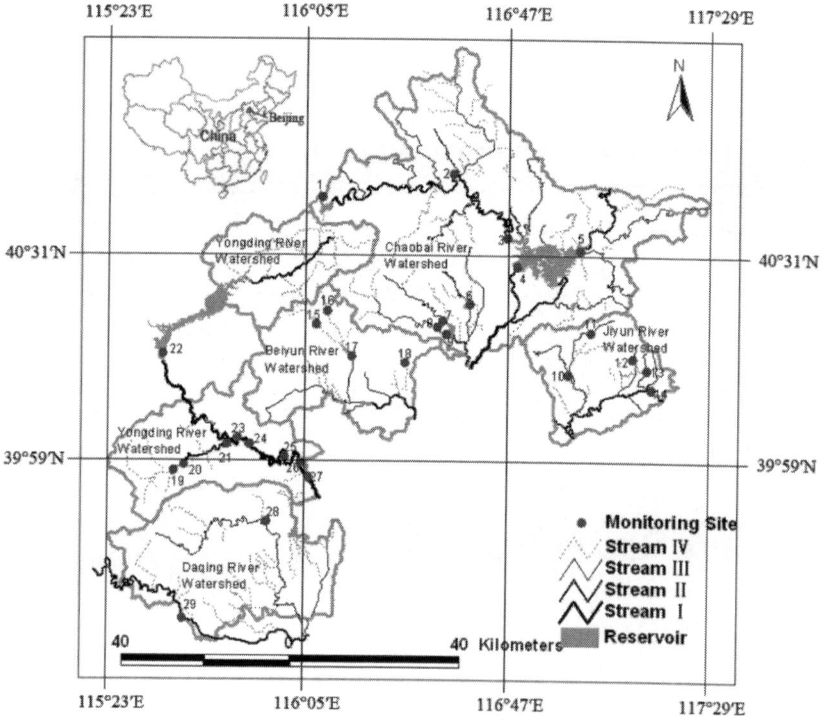

FIGURE 1: Study area and monitoring sites.

With the population increase and urban sprawl of Beijing, mountain agriculture has played increasingly important roles in areas such as services, the economy, ecological security and tourism. Figure 2 shows that the gross value of agricultural output in the study area increased quickly with the pressure for arable land resources in plain areas that have become non-agricultural land owing to city sprawl, particularly in the high development periods of the mid 1990s. The past studies suggested that the increase in agricultural output mainly depended on the input of a large amount of non-renewable resources, especially agricultural chemical fertilizers, pesticides and plastic film, according to the correlation analyses of agricultural inputs and outputs in 2000 [24]. The main non-renewable agricultural inputs have still increased in recent decade years, which has led to greatly increased environmental damage such as water pollution, soil contamination and air pollution caused by agriculture, especially high intensity industrialized agriculture.

1.2.2 SURFACE WATER QUALITY DATA

There are 27 monitoring sites in the study area. The monitoring Sites 1–8, 9–13, 14–17, 18–25 and 26–27 correspond to the Chaobai River, Juyun River, Beiyun River, Yongding River and Daqing River watersheds, respectively (Figure 1). The streams on which they are respectively located are listed in Table 1. Water samples were taken at these stations monthly from May to October 2000. Of these, June, July, August and November in 2000 were the rainy reason, and the rest was the dry season. A total of 162 samples were collected at the 27 sites of the Beijing mountainous areas. Water samples were analyzed in the laboratory for eight water quality characteristics, including nitrate-nitrogen (NO_3-N), permanganate index (COD_{Mn}), biochemical oxygen demand for five days (BOD_5), total nitrogen (TN), total phosphorus (TP), total mercury (Hg), total cadmium (Cd) and total lead (Pb). Then the monthly average values of water quality characteristics for each site were derived, which were used for our statistical assessment.

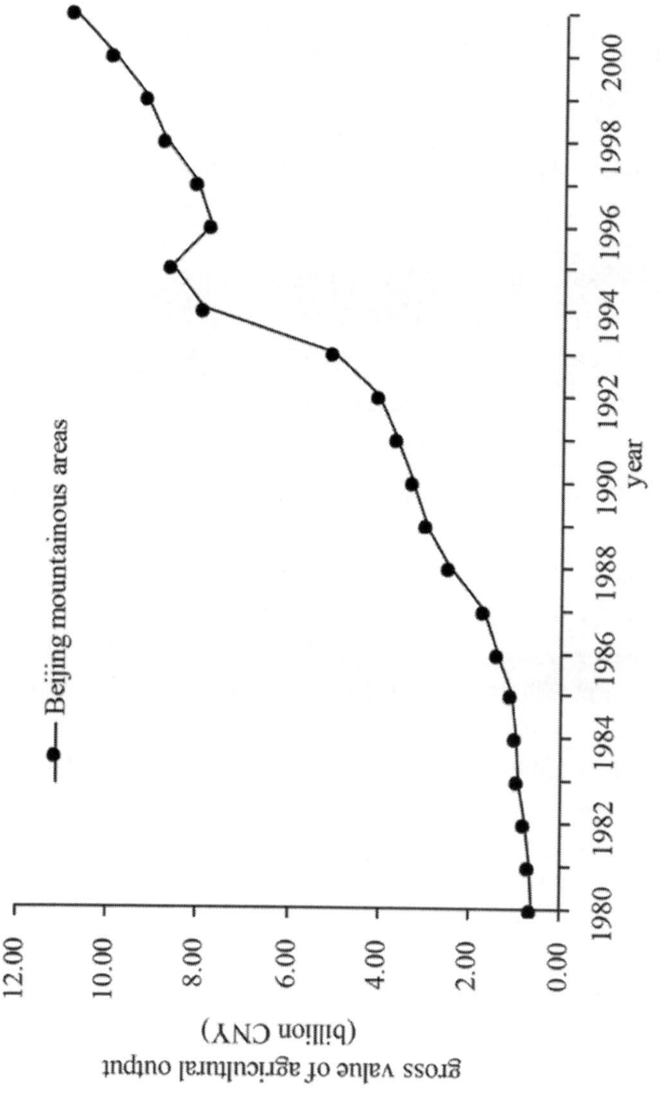

FIGURE 2: Gross agricultural output for the Beijing mountainous areas.

TABLE 1: The streams and watersheds of the 27 monitoring sites.

Watersheds	Stream	Site
Chaobai River	Bai River	1–3
	Chao River	4
	Yanqi River	5
	Huaisha River	6
	Huaijiu River	7
	Huai River	8
Jiyun River	Cuo River	9
	Zhenluoying Rock River	10
	Huangsongyu Rock River	11
	Jiangjunguan Rock River	12
	Ju River	13
Beiyun River	Deshengkou Ditch	14
	Zhuishikou Ditch	15
	Dongsha River	16
	Qintun River	17
Yongding River	Qingshui River	18–20
	Yongding River	21–25
Daqing River	Dashi River	26
	Juma River	27

TABLE 2: Bivariate correlation coefficients of water quality variables.

	NO_3-N	COD_{Mn}	BOD_5	Hg	Cd	Pb	TN	TP
NO_3-N	1							
COD_{Mn}	0.607[b]	1						
BOD_5	0.655	–0.034	1					
Hg	0.809 [b]	0.857 [b]	–0.611	1				
Cd	0.917 [b]	0.640 [b]	–0.742	0.904 [b]	1			
Pb	0.917 [b]	0.640 [b]	–0.742	0.904 [b]	1.000 [b]	1		
TN	0.996 [a]	–0.829	0.587	0.282	0.107	0.107	1	
TP	0.999 [a]	–0.805	0.621	0.242	0.065	0.065	0.999 [a]	1

[a] *Significant at the 0.01 level.* [b] *Significant at the 0.05 level.*

The surface water quality characteristics NO_3-N and COD_{Mn} have been considered as two of the four water pollutant load control indexes in China. Furthermore, the relationship between water quality characteristics was tested at the significance levels of $p < 0.01$ and 0.05, which showed that NO_3-N had a high positive correlation with TN, TP, Hg, Cd and Pb, and CODMn had a positive correlation with Hg, Cd and Pb (Table 2).

Therefore, we focused on the two conventional water quality characteristics NO_3-N and COD_{Mn}, which not only can reduce the complexity of the study, but also is of great significance for water resources management.

1.2.3 MULTI-SCALE WATERSHED DELINEATION

1.2.3.1 BASIC WATERSHED UNITS

A watershed is the up-slope area contributing flow to a given location. Such a feature is also variously referred to as a catchment or basin, and comprises a part of a hierarchy in that any given watershed is generally part of a larger watershed [19]. A digital elevation model (DEM) with 30 m × 30 m grid resolution was adopted to create basic watershed units for the Beijing mountainous area using the hydrologic functions in the ArcView extensions. The minimum number of cells for a stream network in the hydrologic functions is very important for the watershed delineation. Many studies have shown that when the minimum number of cells was smaller, the extracted stream networks were denser, and the created basin areas were smaller. The stream network extracted by hydrologic functions should be similar to that existing in Nature, and each of the monitoring sites should be located at the pour point of different basic watershed units for the purpose of this study. Therefore, the thresholds of 15,000, 12,000, 10,000, 8,000 and 5,000 were chosen as the minimum number of cells for a stream network during the test of basic watershed unit creation. The results indicated that a few monitoring sites were located in the same basic watershed unit when the thresholds was bigger than 10,000, and the stream networks extracted were far denser than natural stream network when the threshold was less than 10,000. Ultimately, the thresholds of 10,000, i.e.,

900 ha, which is far smaller than the average area of the Beijing mountain towns (9,285 ha), was determined as the minimum number of cells for a stream network to delineate basic watershed units (Figure 3).

1.2.3.2 DELINEATING MULTI-SCALE WATERSHEDS AS WATERSHED ANALYSIS UNITS

Multi-scale watersheds corresponding to the 27 monitoring sites were identified as watershed analysis units for further analysis according to the flow direction and rivers, respectively. Since some monitoring sites are located on the same streams, such as Sites 1–3 (Bai River), Sites 18–20 (Qingshui River) and Sites 21–25 (Yongding River), perhaps they are statistically highly correlated and, to some extent, all the upstream points contribute to the measurements of any monitoring point. This is statistically undesirable and would produce strongly biased results. To solve this problem, the independence of these sites' data should be tested. A serial autocorrelation test was adopted to analyze the possible correlation between neighboring sites. The test results showed that these upstream monitoring sites had little contribution to their nearest downstream sites for NO_3-N and COD_{Mn} water quality data in 2000. Therefore, the monitoring samples can be considered independent for further statistical analysis, and the watersheds contributing flow to these sites cannot be included in the their nearest downstream sites' watersheds when delineating the watershed analysis units.

Firstly, the basic watershed units contributing flow to every monitoring site were delineated as the whole watershed of every monitoring site, which did not have a nest relation according to the result of serial autocorrelation test mentioned above (Figure 4). Table 3 shows the number of towns covered by the whole watershed for every monitoring site. Subsequently, multi-scale watersheds were determined in the whole watershed area of every monitoring site. The basic watershed units directly contributing flow to every monitoring site were considered as the first order watershed (Zone 1). The Zone 1 and the basic watershed units directly contributing flow to the Zone 1 were together defined as second order watershed (Zone 2).

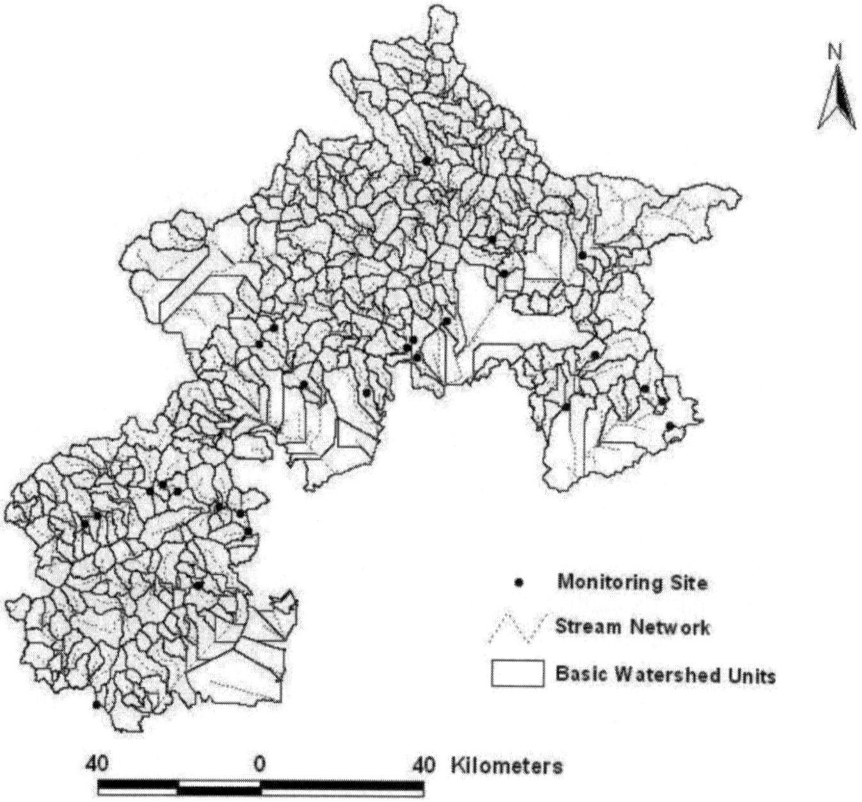

FIGURE 3: Basic watershed units delineated in the study area.

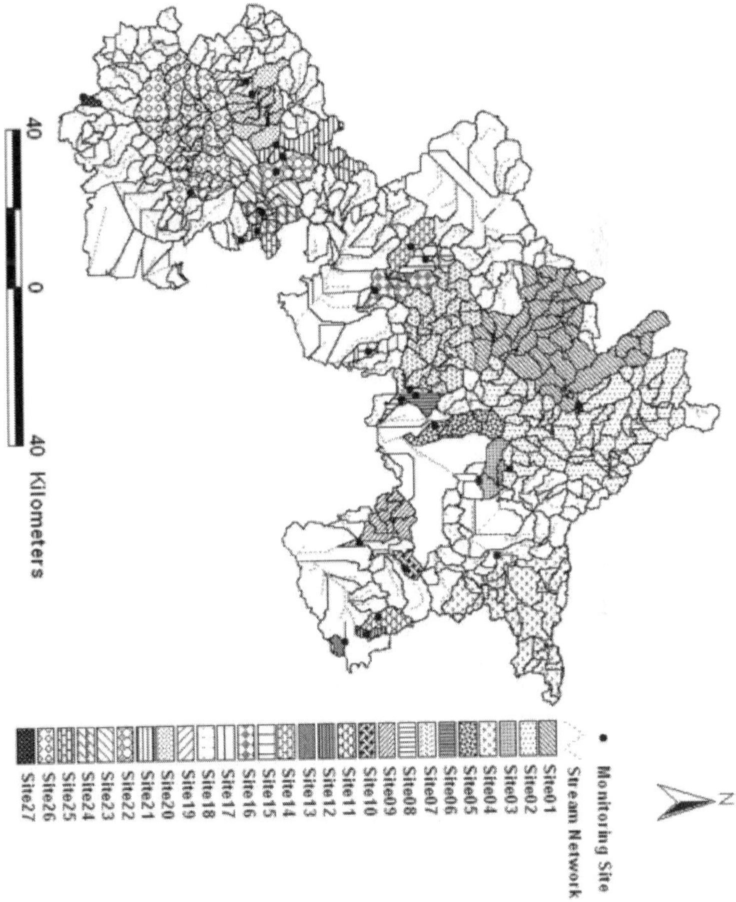

FIGURE 4: The whole watersheds for the 27 monitoring sites.

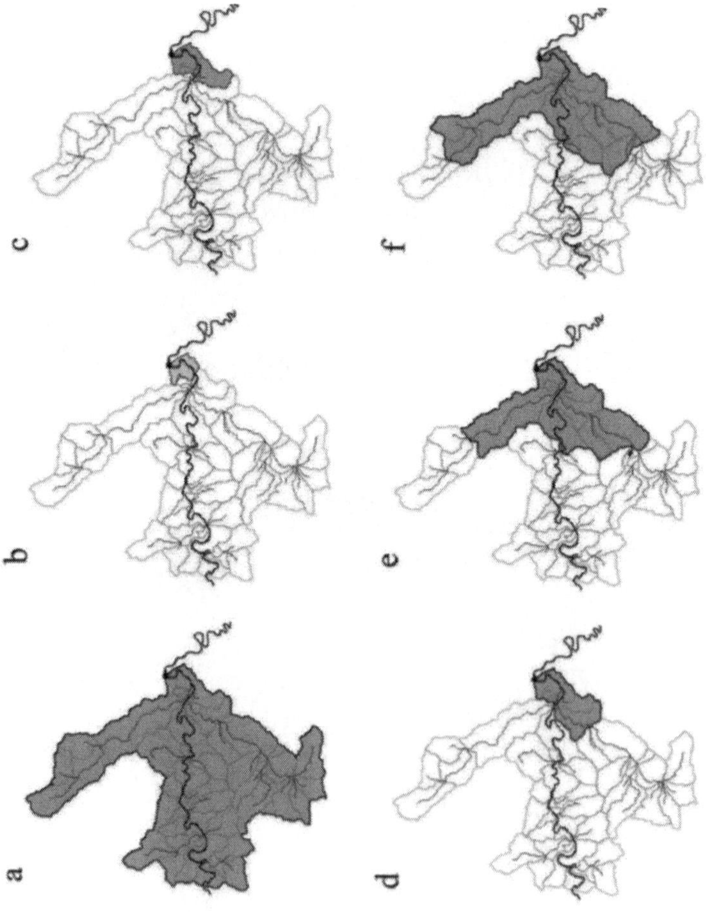

FIGURE 5: Illustration of how the multi-scale watersheds were defined: (a) The whole watershed for Site 1. (b) Definition of the Zone 1. (c) Definition of the Zone 2. (d) Definition of the Zone 3. (e) Definition of the Zone 4. (f) Definition of the Zone 5.

By analogy, the next order watersheds were derived and named in order third order watershed (Zone 3), fourth order watershed (Zone 4) and fifth order watershed (Zone 5), etc. Figure 5 illustrates the process of delineating multi-scale watersheds for Site 1. Table 3 also lists the number of scales for every monitoring site.

TABLE 3: The number of scales and towns be covered for 27 monitoring sites.

Site	Towns	Scale	Site	Towns	Scale
1	6	>10	14	3	1
2	7	>10	15	2	3
3	2	2	16	4	4
4	6	7	17	1	2
5	4	3	18	1	7
6	2	1	19	2	3
7	5	10	20	2	8
8	3	1	21	2	4
9	4	5	22	2	2
10	2	1	23	4	6
11	4	1	24	4	3
12	3	1	25	2	2
13	2	1	26	8	>10
			27	1	1

1.2.4 AGRICULTURAL LAND USE INTENSITY FOR WATERSHED

With more attention being paid to land use and land cover change, an approach to assess agricultural land use intensity including agricultural input and output intensity on a general basis has been developed in our previous work [24]. The measurement and assessment of agricultural land use intensity was preformed at municipal/town level because the agricultural inputs and outputs information derived from census data are aggregated and officially reported at this level, and this level also is the smallest administrative unit as planning and management purpose in China. Four indices of agricultural input intensity and six of output intensity have been derived with principal component analysis at this level for the Beijing mountainous areas using the amount of emergy of each agricultural input and output derived from agricultural census data [24]. Their eigenvectors are given in Tables 4 and 5. The several indices reflecting agricultural input and output intensity were dimensionless, and the meanings of these indices according to their eigenvectors are shown in Table 6. The higher the index value, the greater the agricultural input or output intensity.

TABLE 4: Eigenvectors of the input intensity [24].

Components	IPC_1	IPC_2	IPC_3	IPC_4
Sunlight	0.990	−0.075	−0.083	0.018
Rain, chemical energy	0.983	−0.056	−0.086	0.070
Rain, geopotencial energy	0.932	−0.163	−0.159	−0.096
Earth cycle	0.991	−0.071	−0.086	0.028
Wind, kinetic energy	0.991	−0.071	−0.086	0.028
Soil loss	0.991	−0.071	−0.086	0.028
Agricultural electricity	−0.252	0.364	0.527	0.156
Nitrogen fertilizer	−0.102	0.792	0.165	0.448
Phosphorus fertilizer	−0.062	0.852	0.011	0.305
Potash fertilizer	−0.103	0.890	0.113	0.139
Compound fertilizer	−0.119	0.809	0.270	0.184
Pesticides	0.005	0.109	0.787	−0.103
Agricultural plastic mulch	−0.165	−0.024	0.673	0.180
Machinery power	0.003	0.619	0.348	−0.085
Human labor	0.140	0.344	0.620	0.494
Livestock labor	0.135	0.020	−0.034	0.875
Organic manure	−0.088	0.471	0.268	0.648
Seed	−0.090	0.761	0.234	0.425

TABLE 5: Eigenvectors of the output intensity [24].

Components	OPC_1	OPC_2	OPC_3	OPC_4	OPC_5	OPC_6
Grain crops	0.569	0.423	0.258	0.186	0.055	0.045
Oil crops	−0.021	−0.003	−0.050	0.911	0.056	0.046
Vegetables	0.761	0.136	−0.105	−0.156	−0.028	−0.092
Fruits	−0.077	−0.088	0.867	−0.131	0.088	0.064
Pork	0.682	0.228	0.226	0.325	0.187	−0.033
Beef	0.240	0.792	−0.061	0.063	−0.139	0.096
Mutton	−0.053	0.109	0.069	0.050	0.021	0.967
Fowl	0.359	0.094	0.548	0.389	−0.175	−0.002
Milks	0.097	0.855	−0.026	−0.071	0.146	0.008
Eggs	0.815	0.111	−0.038	0.003	0.055	−0.124
Forest logging	0.062	0.036	0.026	0.048	0.968	0.020
Fish	0.781	−0.237	−0.077	−0.120	−0.101	0.225

TABLE 6: The indices of agricultural input and output intensity.

	Indices at town level	Meanings	Indices at watershed level
Agricultural input intensity	IPC_1	Natural factors, including soil loss, rainfall and sunlight	WI-IPC_1
	IPC_2	Chemical fertilizer, seed and mechanized power	WI-IPC_2
	IPC_3	Pesticides, agricultural plastic mulch, human labor and agricultural electricity	WI-IPC_3
	IPC_4	Organic manure	WI-IPC_4
Agricultural output intensity	OPC_1	Eggs, vegetables, pork and grain crops	WI-OPC_1
	OPC_2	Milks and beef	WI-OPC_2
	OPC_3	Fruits and fowl	WI-OPC_3
	OPC_4	Oil crops	WI-OPC_4
	OPC_5	Forest cut	WI-OPC_5
	OPC_6	Sheep	WI-OPC_6

For watershed-scale analysis, the indices of agricultural input and output intensity should be translated from the municipal/town level to the watershed level. Because there is a spatial incompatibility between the watershed analysis unit and the municipal/town unit, the weighted values with the percentage of town's agricultural land area in the watershed analysis unit were used as the weights to calculate the agricultural land use intensity indices for watershed analysis unit, which is as follows:

$$WUI_i = \sum_{j=1}^{n} \%WT_j \cdot TUI_j \tag{1}$$

where WUI_i are the agricultural input and output indices for the watershed analysis unit i, $\%WT_j$ is the percentage of town j's agricultural land area in a watershed analysis unit i. TUI_j is the agricultural input and output indices for town j. Therefore, the indices for watershed analysis unit i were the weighted values with agricultural land area percentage used to reflect agricultural input and output intensity at every watershed scale.

1.2.5 AGRICULTURAL LAND USE INTENSITY AND WATER QUALITY LINKAGE

The question of a relationship between agricultural land use intensity and water quality was examined at various scales by applying multiple regression techniques considering nutrient concentrations as dependent variables and the agricultural land use intensity as explanatory variables. The functional form of the relationship for each of these scales is as follows:

$$NPS_i = f(WUI_i) \tag{2}$$

where NPS_i is nutrient concentration for monitoring site in question in watershed analysis unit i, WUI_i is equal to the indices of agricultural input and output intensity for watershed analysis unit i.

In previous studies, the concentration of nutrients over an area can be described in the form of an exponential model or a linear model considering the proportions of land use/land cover as explanatory variables. Delivery of non-point source pollutants from discrete upstream contributing zones to a particular downstream point is a multi-step, often episodic, process [25]. A first order rate equation can be used for modeling nutrient attenuation in flow through various land uses to the nearest stream [25]. Since agricultural land use intensity has been one of the most significant forms of land cover modification, the exponential model was chosen for this research to explore the relationship between agricultural land use intensity and water quality, and to recognize how agricultural land use intensity affects water quality at various watershed scales. Stepwise multiple regression analysis was performed using log transformed dependent variables to reduce the asymmetric distribution of the data. The numbers of scales for 27 monitoring sites were different, thus, the used monitoring sites and sample number at various scales were different in the multi-scale regression analyses (Table 7). Because the sample numbers were too small to make valuable regression analysis above the scale of Zone 5, the multi-scale regression analyses were carried out with SPSS 13.0 at the scales from Zone 1 to Zone 5. Based on this statistical analysis, watershed adaptive response units for each water quality variable can also be identified.

TABLE 7: The monitoring sites at various scale level in the multi-scale analyses.

Scale	Sites	Number
Zone 1	1, 2, 3, 4, 5, 6, 7, 8, 9, 10, 11, 12, 13, 14, 15, 16, 17, 18, 19, 20, 21, 22, 23, 24, 25, 26, 27	27
Zone 2	1, 2, 3, 4, 5, 7, 9, 15, 16, 19, 20, 21, 22, 23, 24, 25, 26	17
Zone 3	1, 2, 4, 5, 7, 9, 15, 16, 19, 20, 21, 23, 24, 26	14
Zone 4	1, 2, 4, 7, 9, 16, 20, 21, 23, 26	10
Zone 5	1, 2, 4, 7, 9, 20, 23, 26	8
Zone 6	1, 2, 4, 7, 20, 23, 26	7
Zone 7	1, 2, 4, 7, 20, 26	6
>Zone 7		<5

1.3 RESULTS

1.3.1 WATER QUALITY ASSESSMENT

Table 8 shows the 2000 yearly mean concentration of NO_3-N and COD_{Mn}. The concentration of NO_3-N in the samples ranged from 0.46 to 12 mg/L, while COD_{Mn} ranged from 1 to 7.4 mg/L. Only the concentration of NO_3-N for Site 12 exceeded to 10 mg/L, which is the Chinese surface drinking water standard limit. Different from NO_3-N, there are five types of surface water environmental quality standard for COD_{Mn}. Since type IV and V cannot be acceptable for drinking water, the type III recommended value of 6 mg/L is considered as the surface drinking water standard limit for COD_{Mn}. According to this standard limit of 6 mg/L, Site 12 was faced with organic contaminant and reducible inorganic substance drinking water pollution. Since Sites 10, 11, 17 and 22 were close to the standard limit of 6 mg/L, especially Site 10, they also had a pollution risk caused by COD_{Mn}.

1.3.2 LINKAGE MODEL RESULTS AT MULTIPLE SCALES

The regression models of NO_3-N and COD_{Mn} at the various watershed scales are shown in Table 9 and Table 10, respectively. The log-transformed NO_3-N exhibited a relationship with WI-IPC_1 and WI-IPC_2 at the scale of Zone 1, while no statistically significant relationships were found between agricultural land use intensity and nitrate level at the other watershed scales. The regression equation in the Zone 1 model, with the value 0.374 of R^2 and the level 0.004 of statistical significance, is as follows:

$$\ln(NO_3\text{-N}) = -0.026 - 0.901\text{WI-IPC}_1 + 0.418\text{WI-IPC}_2 \quad (3)$$

TABLE 8: The concentration of NO_3-N and COD_{Mn} in stream water of 27 monitoring sites.

Monitoring site	NO_3-N			COD_{Mn}		
	NO_3-N (mg/L)	Standard limit	Type	COD_{Mn} (mg/L)	Standard limit	Type
1	1.6	10	Not exceeding	2.4	4	II
2	1.26	10	Not exceeding	2.2	4	II
3	0.79	10	Not exceeding	2.3	4	II
4	3.14	10	Not exceeding	2.1	4	II
5	0.46	10	Not exceeding	3.2	4	II
6	1.67	10	Not exceeding	1.9	2	I
7	2.69	10	Not exceeding	1.5	2	I
8	0.66	10	Not exceeding	2.5	4	II
9	1.95	10	Not exceeding	1.5	2	I
10	3.4	10	Not exceeding	6	6	III *
11	2.57	10	Not exceeding	4.4	6	III *
12	12	10	Exceeding *	7.4	10	IV **
13	1.09	10	Not exceeding	2.6	4	II
14	0.86	10	Not exceeding	1.4	2	I
15	1.06	10	Not exceeding	1.5	2	I
16	0.37	10	Not exceeding	2.7	4	II
17	0.18	10	Not exceeding	4.9	6	III *
18	1.68	10	Not exceeding	1.4	2	I
19	1.72	10	Not exceeding	3.2	4	II
20	1.78	10	Not exceeding	1.3	2	I
21	1.88	10	Not exceeding	2.1	4	II
22	1.6	10	Not exceeding	4.4	6	III *
23	1.33	10	Not exceeding	4	4	II
24	1.51	10	Not exceeding	3.9	4	II
25	1.51	10	Not exceeding	4	4	II
26	3.51	10	Not exceeding	1	2	I
27	1.76	10	Not exceeding	1.6	2	I

Although the coefficients of determination R^2 is relatively weak, the model is statistically significant. The model suggests that natural factors act as sinks, and as the input intensity of natural factors including rainfall and sunlight inside the contributing zone (Zone 1) increases, NO_3-N levels downstream decrease. In addition, several artificial inputs, especially chemical fertilizer input, are identified as the second largest contributors of NO_3-N, and as chemical fertilizer input intensity within the contributing zone (Zone 1) increases, NO_3-N levels downstream also increase.

TABLE 9: Multiple regression model of NO_3-N.

Scales	Variable in equation	Standardized Beta	R^2	Sig.	Number of samples
Zone1	WI-IPC_1	−0.469	0.374	0.004	27
	WI-IPC_2	0.412			
Zone2		no variables were entered			17
Zone3		no variables were entered			14
Zone4		no variables were entered			10
Zone5		no variables were entered			8

TABLE 10: Multiple regression model of COD_{Mn}.

Scales	Variable in equation	Standardized Beta	R^2	Sig.	Number of samples
Zone1		no variables were entered			27
Zone2		no variables were entered			17
Zone3		no variables were entered			14
Zone4		no variables were entered			10
Zone5	WI-IPC_3	0.527	0.452	0.001	8
	WI-IPC_4	0.085			

For the water quality index COD_{Mn}, no variables were entered in the stepwise regression analysis for the scales from Zone 1 to Zone 4, while the most important explanatory variables were the WI-IPC_3 and WI-IPC_4

at the watershed scales of Zone 5. The regression equation between agricultural land use intensity and permanganate index in the Zone 5 model, with the value 0.452 of R^2 and the level 0.001 of statistical significance, is as follows:

$$\ln(COD_{Mn}) = 0.745 + 0.514WI\text{-}IPC_3 + 0.052WI\text{-}IPC_4 \quad (4)$$

In the regression model of permanganate index, the two input intensities of pesticides and agricultural plastic mulch and organic manure inside the contributing zone (Zone 5) both have the positive impact on the permanganate level downstream. Therefore, the input of pesticides and agricultural plastic mulch is considered as the larger contributor than the organic manure input.

1.4 DISCUSSION AND CONCLUSIONS

1.4.1 AGRICULTURAL INPUT INTENSITY AND SURFACE WATER QUALITY RISK

Land use/land cover management, particularly high-input agriculture, is considered to be an important source of pollution export from catchments and frequently has been identified as a major contributor of surface water pollution [26]. The above results and analysis provide insight into the linkages between agricultural land use intensity and regional surface water quality. For the Beijing mountainous study area, several groups of agricultural input affecting surface water quality were identified during the year 2000. The results indicated that the explanatory variables behind the various water quality indexes were quite different at the respective significant watershed scales. The view that nitrate may be a useful general indicator of intensive land use was supported by previous work by Hunt et al [27]. As in Hunt [27], nitrate in particular can be considered as a useful indicator of intensive natural factors and agricultural chemical fertilizer input at the significant watershed scale in the Beijing mountainous areas. The input intensity of pesticide and agricultural plastic mulch in the Beijing mountainous areas watersheds has increased drastically between 1984 and 2000, which resulted in the permanganate index pollution risk. Several studies have indicated that the proportion of vegetable-planted land exhibited a positive correlation with permanganate index [28]. For the Beijing mountainous areas, vegetable outputs depended principally on the abundant inputs of pesticide and agricultural plastic mulch in 2000, according to our previous research [24].

1.4.2 WATERSHED ADAPTIVE RESPONSE UNIT

The significant scales at which there were statistically significant relationships between agricultural land use intensity and each water quality variable were identified on the basis of the multi-scale regression analysis, which were considered as the watershed adaptive response units for each water quality variable. Thus, the first order watershed (Zone 1) of 27 monitoring points was the adaptive response unit for nitrate-nitrogen, while the fifth order watershed (Zone 5) was the adaptive response unit for the permanganate index.

In the Beijing mountainous study area, watershed adaptive response units differ with the water quality variables being assessed, which are related with transformation regularity of nitrate-nitrogen and permanganate index. After the use of nitrate fertilizer that is the source of nitrate-nitrogen on agricultural fields, nitrate-nitrogen formed by nitrification is either absorbed and utilized by plants or transformed into gaseous nitrogen through denitrification under reducing conditions. Therefore, only the agricultural inputs inside the first order watershed zones can make a significant contribution to the concentration of nitrate-nitrogen at the pour-point, while that inside other order watershed zones has little influence on the nitrate-nitrogen level at the pour-point with the action of long distance transport.

Compared to nitrate-nitrogen, permanganate index contamination is relatively steady. As the main source, the transformation time of agricultural plastic mulch and pesticides is relatively long. It could not make a significant contribution to the concentration of permanganate index until it accumulates. Therefore, the smaller area basins, such as first order watershed to the fourth order watershed, hardly respond to permanganate index as a contributing zone.

The definition of watershed adaptive response unit based on the basic watershed units, actually a contributing zone, is very meaningful for the purpose of more effective watershed management. It is important to address the fine-scale management issues relate to watershed adaptive response units for every water quality parameter. The adjustment of agricultural input structure and intensity may be carried out inside the individual watershed adaptive response units.

1.4.3 MODELING

The multiple linear regression model performed using log transformed dependent variables, which was adopted in many previous studies to explore the relationship between land use and stream water, can also provide insight into the linkages between agricultural land use intensity and stream water quality at multiple watershed scales. The statistical models in this study are valuable in examining the relative sensitivity of water quality indexes to alterations in agricultural land use intensity inside the various contributing zones when coupled with expert knowledge. The modeling results can also further help to identify the cause-and-effect relationships between agricultural input intensity and stream water quality inside the watershed adaptive response units, which are important in the management of water quality. The modeling, although statistically significant, showed the relatively weak coefficients of determination. It may be that the spatial incompatibility between the watershed spatial unit and the municipal/town unit was actually existed, or that other potential factors influencing stream water quality variable were not included in the analysis. All of these are worthy of further research.

Although multiple linear regression models are an effective approach for identifying significant agricultural input intensity affecting water quality and explaining the relationship between agricultural land use intensity and stream water quality, they do not appear to quantitatively estimate contribution of respective agricultural land use intensity on the water quality because they are only based on the existence of statistical significance in the analysis data, rather than any mechanistic relationships between sources and receptors. Our future research will focus on understanding the exact mechanisms of the effect of agricultural land use intensity on stream water quality by adopting an alternative—sources-receptors model based on the mass balance approach.

REFERENCES

1. Lambin, E.F.; Rounsevell, M.D.A.; Geist, H.J. Are agricultural land-use models able to predict changes in land-use intensity? Agr. Ecosyst. Environ. 2000, 127, 321–331.
2. Tilman, D. Global environmental impacts of agricultural expansion: The need for sustainable and efficient practices. Proc. Nat. Acad. Sci. USA 1999, 96, 5995–6000.
3. Bouwman, A.F.; Booij, H. Global use and trade of feedstuffs and consequences for the nitrogen cycle. Nutr. Cycl. Agroecosys. 1998, 52, 261–267.
4. Medema, G.J.; van Asperen, I.A.; Havelaar, A.H. Assessment of the exposure of swimmers to microbiological contaminants in fresh waters. Water Sci. Technol. 1997, 35, 157–163.
5. Tong, S.T.Y.; Chen, W. Modeling the relationship between land use and surface water quality. J. Environ. Manage. 2002, 66, 377–393.
6. Ribolzi, O.; Cuny, J.; Sengsoulichanh, P.; Mousquès, C.; Soulileuth, B.; Pierret, A.; Huon, S.; Sengtaheuanghoung, O. Land use and water quality along a Mekong Tributary in Northern Lao PDR. Environ. Manage. 2011, 47, 291–302.
7. Seeboonruang, U. A statistical assessment of the impact of land uses on surface water quality indexes. J. Environ. Manage. 2012, 101, 134–142.
8. Chaplot, V.; Saleh, A.; Jaynes, D.B.; Arnold, J. Predicting water, sediment and NO3-N loads under scenarios of land-use and management practices in a flat watershed. Water Air Soil Pollut. 2004, 154, 271–293.
9. Cao, W.; Bowden, W.B.; Davie, T.; Fenemor, A. Modelling impacts of land cover change on critical water resources in the Motueka River catchment, New Zealand. Water Resour. Manag. 2009, 23, 137–151.
10. Bhattarai, G.; Srivastava, P.; Marzen, L.; Hite, D.; Hatch, U. Assessment of economic and water quality impacts of land use change using a simple bioeconomic model. Environ. Manage. 2008, 42, 122–131.
11. Rhodes, A.L.; Newton, R.M.; Pufall, A. Influences of land use on water quality of a diverse New England watershed. Environ. Sci. Technol. 2001, 35, 3640–3645.
12. Li, S.; Gu, S.; Tan, X.; Zhang, Q. Water quality in the upper Han River basin, China: The impacts of land use/land cover in riparian buffer zone. J. Hazard. Mater. 2009, 165, 317–324.
13. Mattikalli, N.M.; Richards, K.S. Estimation of surface water quality changes in response to land use change: Application of the export coefficient model using remote sensing and geographical information system. J. Environ. Manage. 1996, 48, 263–282.

14. Xiao, H.; Ji, W. Relating landscape characteristics to non-point source pollution in mine waste-located watersheds using geospatial techniques. J. Environ. Manage. 2007, 82, 111–119.
15. Wang, X. Integrating water-quality management and land-use planning in a watershed context. J. Environ. Manage. 2001, 61, 25–36.
16. Bahar, M.M.; Ohmori, H.; Yamamuro, M. Relationship between river water quality and land use in a small river basin running through the urbanizing area of Central Japan. Limnology 2008, 9, 19–26.
17. Kang, J.H.; Lee, S.W.; Cho, K.H.; Ki, S.J.; Cha, S.M.; Kim, J.H. Linking land-use type and stream water quality using spatial data of fecal indicator bacteria and heavy metals in the Yeongsan river basin. Water Res. 2010, 44, 4143–4157.
18. Liu, Z.; Li, Y.; Li, Z. Surface water quality and land use in Wisconsin, USA—A GIS approach. J. Integr. Environ. Sci. 2009, 6, 69–89.
19. Basnyat, P.; Teeter, L.D.; Lockaby, B.G.; Flynn, K.M. The use of remote sensing and GIS in watershed level analyses of non-point source pollution problems. For. Ecol. Manage. 2000, 128, 65–73.
20. Sliva, L.; Williams, D.D. Buffer zone versus whole catchment approaches to studying land use impact on river water quality. Water Res. 2001, 35, 3462–3472.
21. Maillard, P.; Pinheiro Santos, N.A. A spatial-statistical approach for modeling the effect of non-point source pollution on different water quality parameters in the Velhas river watershed-Brazil. J. Environ. Manage. 2008, 86, 158–170.
22. Jung, K.W.; Lee, S.W.; Hwang, H.S.; Jang, J.H. The effects of spatial variability of land use on stream water quality in a costal watershed. Paddy Water Environ. 2008, 6, 275–284.
23. Wu, Q.; Li, H.Q.; Wang, R.S.; Paulussen, J.; He, Y.; Wang, M.; Wang, B.H.; Wang, Z. Monitoring and predicting land use changes in Beijing using remote sensing and GIS. Landsc. Urban Plan 2006, 78, 322–333.
24. Zhang, W.W.; Li, H.; Huo, X.N.; Sun, D.F.; Zhou, L.D. Agricultural land use intensity based on emergy analysis (in Chinese, with English abstract). Trans. CSAE 2009, 25, 204–210.
25. Phillips, J.D. Evaluation of North Carolina's estuarine shoreline area of environmental concern from a water quality perspective. Coast. Manage. 1989, 17, 103–117.
26. Hoorman, J.; Hone, T.; Sudman, T.; Dirksen, T.; Iles, J.; Islam, K.R. Agricultural impacts on lake and stream water quality in Grand Lake St. Marys, Western Ohio. Water Air Soil Pollut. 2008, 193, 309–322.
27. Hunt, J.W.; Anderson, B.S.; Phillips, B.M.; Tjeerdema, R.S.; Richard, N.; Connor, V.; Worcester, K.; Angelo, M.; Bern, A.; Fulfrost, B.; et al. Spatial relationships between water quality and pesticide application rates in agricultural watersheds. Environ. Monit. Assess. 2006, 121, 243–260.
28. Li, S.; Gu, S.; Liu, W.; Han, H.; Zhang, Q. Water quality in relation to land use and land cover in the upper Han River Basin, China. Catena 2008, 75, 216–222.

CHAPTER 2

DISCUSSION ON SUSTAINABLE WATER TECHNOLOGIES FOR PERI-URBAN AREAS OF MEXICO CITY: BALANCING URBANIZATION AND ENVIRONMENTAL CONSERVATION

TIEMEN A. NANNINGA, IEMKE BISSCHOPS, EDUARDO LÓPEZ, JOSÉ LUIS MARTHNEZ-RUIZ, DANIEL MURILLO, LAURA ESSL, AND MARKUS STARKL

2.1 INTRODUCTION

In urban and densely populated areas, centralized solutions for water management are usually favored over small scale solutions based on principles of economy of scale and economy of density [1–3]. In this traditional approach, water, wastewater and solid waste are often transported over large distances. These centralized water supply, sanitation and solid waste solutions often struggle to keep up with rapid expansion of large cities in the South, such as Mexico City, not only due to unregulated and uncontrolled urbanization, but also due to high construction costs [4]. Indeed, it is in these peri-urban areas that there is a trend towards providing water supply, sanitation and solid waste management technologies on a decentralized,

This chapter was originally published under the Creative Commons Attribution License. Nanninga TA, Bisschops I, López E, Martínez-Ruiz JL, Murillo D, Essl L, and Starkl M. Discussion on Sustainable Water Technologies for Peri-Urban Areas of Mexico City: Balancing Urbanization and Environmental Conservation. Water ***4*** *(2012); 739-758. doi:10.3390/w4030739.*

on-site scale [4–8]. However, stakeholders may prefer centralized solutions because of the convenience of a centralized system.

The introduction of decentralized technologies allows for the development of new opportunities that enable the recovery and reuse of resources in the form of water, nutrients and energy. This resource-oriented management of water, nutrients and energy requires a sustainable system aimed at low resource use and high recovery and reuse rates.

Instead of designing each sector separately, which has been traditionally done, this article proposes and discusses a concept that combines the in- and outflows of the different sectors, reusing water and other liberated resources where possible, which is illustrated in Figure 1. "Joints" for possible integration of these sectors are: water reuse, nutrient recycling and energy recovery. In the decentralized concept all wastewater in a region is seen as a separate water, nutrient and energy source, and is evaluated for its suitability as a water source for a specific use, such as agriculture, nonpotable domestic purposes or forest irrigation. With letting the final users determine the quantitative and qualitative water requirements needed for a specific purpose such as agriculture, industry or (nonpotable) household activities, the user or purpose of the wastewater determines the quality to which the wastewater is improved, which is the opposite of the conventional water chain where governmental legislation or permits dictate the treatment level. This reverse water chain [9–11], results in a new and novel view of waste and wastewater treatment and water sourcing where legislation should allow for tailor-made solutions. This may also reduce costs as not all waste and wastewater need to be treated up to high standards, since a part can also be used for purposes only requiring lower standards.

This paper shows and demonstrates examples of different types of sustainable water technologies that can be implemented in the peri-urban areas of Mexico City. An innovative participatory planning method using a scenario building methodology at project level was applied. This method has allowed for a meaningful and intensive involvement of a variety of key stakeholders in the planning process. It also helped to identify the opportunities and limitations of resource oriented water technologies within an urban planning context highlighting the conflict between environmental conservation and urbanization.

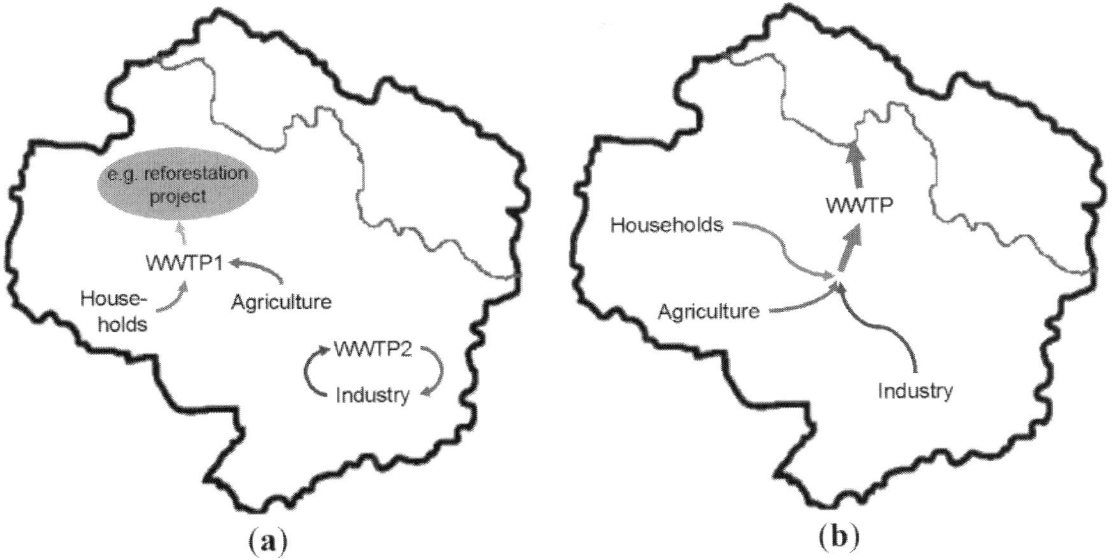

FIGURE 1: (a) Traditional way of thinking about wastewater treatment; and (b) resource oriented way of thinking on a regional scale. WWTP1 and WWTP2 can constitute completely different technologies (WWTP = Wastewater Treatment Plant).

2.2 CASE STUDY

The case study is located in Xochimilco, one of the 16 boroughs (delegaciones) in Mexico City. The entire delegación Xochimilco has an area of approximately 12.5×10^7 m^2 [12,13], of which 2.5×107 m^2 are urbanized. The peri-urban areas (which can be legal or illegally urbanized) cover the additional 3.0×10^7 m^2, and serve for this study (Figure 2). These peri-urban areas have a mixed topography with plain (1.26×10^7 m^2) as well as mountainous areas (1.74×10^7 m^2). Xochimilco has a high biological and cultural value due to the wetlands and chinampas. Chinampas are rectangular land plots (usually 10 by 100 meters) surrounded by narrow canals used for irrigation, fishing, transportation and as a water source. The construction of chinampas dates from pre-colonial times and they are used for agricultural purposes [12–15]. In general, they are considered to be biologically very diverse and agriculturally very high productive areas [12,16,17]. Farmers cultivating crops on a chinampa are called chinamperos. It was listed as a World Heritage site in 1986 and as a wetland of international importance in 2004 [12,18].

The trends of the past four decades confirm a dual-pressure process in Xochimilco, referring to progressive shortages of water and expansion of other land uses. Natural systems (water interactions, soil, vegetation, fauna) are under pressure by the urbanization. In recent years, water has been injected into the aquifer and is used as the main water supply source to retard groundwater depletion of Mexico City. The effluent of a wastewater treatment plant, Cerro de la Estrella, is discharged to the surface water bodies of Xochimilco to ensure that water remains [12]; if this is not done, the surface water bodies will dry as the natural springs have been depleted [19]. Nevertheless, there are no signs that the drying effect on both groundwater and surface water can be countered, not to mention the environmental pollution that takes place due to the discharge of untreated wastewater by households living near or on the chinampas. If this trend continues, the expected scenario for the area is that there will be a large water deficit in the next century [20]. The peri-urban zone has been experiencing problems with services for water, wastewater and solid waste management. This situation will be aggravated due to the unregulated urbanization, definitely.

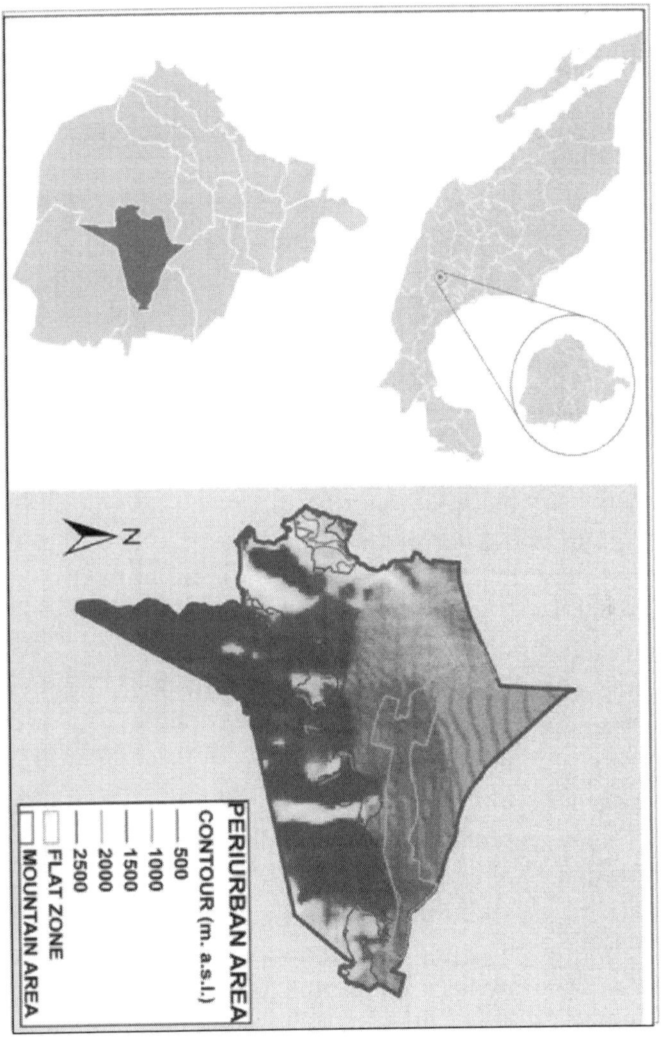

FIGURE 2: Case study area of Mexico City, consisting of the peri-urban zones of Xochimilco.

2.3 METHODOLOGY

2.3.1 PARTICIPATORY PLANNING

Participatory planning is considered an important aspect for achieving sustainable water services, e.g., [21–23]. In this project an innovative approach using scenario building methodology was applied. From the wide range of available methods for scenario building, in this project we were in particular interested in those which allow the users to participate in shaping the development of their region. An example for such a method is the Future Workshop (FW) method. This method allows participants to become involved in creating their preferred future. A "classic" FW, according to Jungk and Müller [24], consists of five phases:

1. The preparation phase;
2. The critique phase;
3. The fantasy phase;
4. The implementation phase;
5. The follow-up phase.

In this study, the scenario workshops conducted under this study encompassed the first three phases, but can be adapted to the local needs and situation as follows:

1. Presentation of the overall study;
2. Presentation of urban problems in the case study area;
3. Presentation of trend analysis study;
4. Participants wrote what they like and what they do not like about the current situation on cards;
5. Elaboration of vision "Xochimilco 2030" in three mixed working groups;
6. Presentation of visions.

The scenario workshop aimed at the identification of different options for future regional development. Building on the outcomes of the scenario workshops, a workshop for participatory planning was conducted. This workshop focused more on the technical aspects with respect to water, wastewater and solid waste management. It encompassed two main phases: the existing environmental problems in the area were discussed; and the participants identified possible solutions and highlighted the main conflicts and barriers that need to be overcome to implement those solutions.

2.3.2 DEVELOPMENT OF CONCEPT SCENARIOS

The results of the scenario building and the participatory planning were combined and then different concept scenarios were developed. A concept scenario encompassed a coherent set of water technologies that are suitable for a different future development scenario (e.g., urbanization or conservation). The concept scenarios were then furnished with a set of suitable technologies.

2.3.3 TECHNICAL FEASIBILITY STUDY

Next, a feasibility was conducted which aimed at demonstrating the technical feasibility of the identified technologies for each concept scenario. As a detailed feasibility study for the entire case study area was beyond the scope of this study, a smaller area was considered much better to suit for "testing" the concepts and its technologies. For the selection, some criteria including infrastructure, urbanization, remoteness and socio-economic conditions were applied to ensure the selection is representative for most peri-urban area in Xochimilco. This included one small area in the lowland chinampa area (La Conchita) and another in the mountainous areas (San Martín). The detailed feasibility study was then conducted for each concept scenario in the selected smaller areas. The following tasks were conducted:

1. A detailed survey of the existing infrastructure in the case study area and a household level survey in the selected smaller areas.
2. A detailed technical feasibility study, which included technical design and drawings of the set of technologies within each concept scenario, thus demonstrating their technical feasibility.

2.3.4 RESOURCES FLOW ANALYSIS

The flows of the different resources were analyzed qualitatively and mapped out for each concept scenario. The resources analyzed were water, Nitrogen (N), Phosphorus (P), organic matter (OM) and energy. Then for selected technologies calculations were performed to quantify the flows. The input data used for these calculations are shown in the Tables 1–3 below:

TABLE 1: Amounts and composition of urine and feces [25].

Waste stream	Amount	N (g/L)	P (g/L)	K (g/L)
Urine	1.2 l/p.d	3	0.8	1.3
Feces	150 g/p.d	2	0.6	0.6

TABLE 2: Amount of organic solid waste and general characteristics of compost [26–28].

Waste stream	Amount	Density	N (g/L)	P (g/L)	OM (g/L)
Organic solid waste produced in Xochimilco	650 g/p.d	-	n/a	n/a	n/a
Compost	-	450 kg/m^3	3.2	0.6	120*

* *Calculated from [27,28].*

The analysis and calculations not only show how (local) finite resources such as water, P and energy are recovered from waste (water), but also how they can be reused. N and OM give a clear indication of how pollution of surface water bodies can be lessened and, with P, how valuable nutrients can be recovered and reused. The analysis of the resources was grouped in water, nutrients and energy, and focused on the import and export resources as well as the generation, recovery and usage of resources within the small case study areas.

TABLE 3: Precipitation data (mm) for San Gregorio Atlapulco weather station, Xochimilco [29].

Precipitation	Jan.	Feb.	Mar.	Apr.	May	Jun.	Jul.	Aug.	Sep.	Oct.	Nov.	Dec.
Monthly sum*	13.4	14	11.7	30.7	78.6	147.9	172.2	148.7	111.4	64.2	7.8	6.6
Daily maximum	32.0	20.3	17.8	23.9	36.8	46.8	49	63.4	36.5	42.0	17.5	14.5

* *Monthly sum are according to climatological normals (1971–2000) and daily maximum are observed.*

2.4 RESULTS

2.4.1 RESULTS—PARTICIPATORY PLANNING

With the past and current systems in Xochimilco, natural resources have been using wastefully, continuously. The case of Xochimilco is critical for the future of whole Mexico City as it serves as a strategic water reserve for the city and for aquifer recharge. As shown in the introduction there is a strong conflict between urbanization of Xochimilco and environmental conservation, a scenario workshop was conducted regarding this background. The participants were from different stakeholder groups including chinamperos, local government representatives, NGOs and academics. The visions of the participants were similar, but could not be summarized in one single concept. Conservation of identity, which was very often mentioned during the workshop, can be partly reached by preserving water resources as they are very important part of the local culture of the chinamperos. Economy, society and the environment should be the vertices of a triangle, and have equal value. In the center of that figure may be awareness, culture and education. They could be starting points for technologies that consider also the tradition of the area and constitute an identity and culture focusing on social justice and a balanced environment applying criteria of sustainability. The engine of these processes should be presence of and pressure from citizens: promoting the creation of new spaces and forms of citizenship or re-elaborate some of the dynamics forms of identity and space. Community-based initiatives can be more effective than public policies supported by governmental institutions, unable to sustain over several administrations.

A synthesis of the visions has shown two possible development scenarios (a development scenario is a vision of the users and stakeholders that participated in the Future Workshop on the future development of the area):

1. Conservation of local identity: as this aspect was mentioned very often during the workshop, a scenario which strengthens the local community was developed. It envisages local solutions and some level of engagement with the city government where inhabitants of Xochimilco are involved and participating in decision making processes concerning resource management. Water protection is a crucial component as water is highly connected to the local identity and the traditional form of cultivation in chinampas.
2. Economic development: This scenario builds on economic development. Originally the focus was on touristic development; however, a later refinement to the case study areas has put emphasis on agricultural development. Although the productivity of the chinampas is generally high [12], not all are used for agricultural purposes. Instead, some parts are fallow and others are used for urbanization. By protecting the chinampas from urbanization and promoting agricultural activities through capacity building and for instance new high-value crops income can increase. In addition, Xochimilco has a touristic area that could be promoted and improved. Prerequisites are a functioning water supply and sewage net, a solid waste management system and the protection of biodiversity and the natural environment that makes the area for tourists an attractive, clean destination. Local products could be certified and sold in supermarkets in Mexico City to generate higher income.

Then, a planning workshop that focused on the technical aspects related to water, wastewater and solid waste management was conducted. During the workshop participants defined various problems related to the water supply, wastewater, agriculture and solid waste for peri-urban areas in Xochimilco that they experience or perceive.

Participants acknowledge the water supply problems, including depletion of the aquifer and poor water supply services. In addition, they also indicate that the water quality, both of the environmental water as well as the domestic water, is poor, which induces several health problems. The participants perceive the discharge of poorly or un-treated wastewater to the canals, streets and environment as the main problem. This results in water pollution, loss of biodiversity and health problems. The performance of the wastewater treatment plants is perceived as being poor, both in terms of quantity and quality. The main agricultural problems are related to a lack of knowledge of farmers and the loss of agricultural land and indigenous agricultural practices such as the chinampas. The absence of solid waste classification, collection and recycling is perceived as a problem, which is mainly related to the lack of awareness about the possibilities for separating and recycling solid waste. Participants have formulated possible technical solutions to the above mentioned problems. They proposed to introduce alternative technologies for water supply, such as rainwater harvesting and groundwater recharge systems. Filters for grey water and the installation of dry toilets could improve the situation concerning untreated wastewater discharge. For solid waste management, the initiation of a waste separation program was proposed by the participants.

2.4.2 RESULTS—CONCEPT SCENARIOS

Each concept scenario is related to one of the possible development scenarios and encompasses a set of technologies that is suitable for the development scenario. In addition, a new concept scenario has been created that is related to increased urbanization. Even if that was not preferred by the participants of the scenario workshop it is a very realistic scenario as a recent trend analysis based on GIS data has shown (see Figure 3). The identified three concept scenarios are summarized in Table 4.

The proposed technologies for concept scenarios 1 and 2 have a strong focus on the reuse and recycling of water, nutrients and energy. Table 5 shows examples for technologies that allow a recovery of nutrients, energy and reuse of water.

TABLE 4: Key objectives and characteristics of the three concept scenarios.

Scenario	Key objectives	Characteristics of technologies under this scenario
1. Local identity	The goal of this scenario is the conservation of local identity which is related to the cultivation of chinampas and the prevention of external influences.	In this concept scenario individual technical solution are preferred over centralized ones to become more independent from Mexico City.
2. Economic development	The goal of this scenario is economic development with a strong focus on agriculture. In the mountainous areas where no agriculture is practiced, there is a focus on community development.	In this scenario there is a strong emphasis on sanitation systems that allow the reuse of nutrients and the water in the chinampas or in some other areas to improve the agricultural production. In the hilly area, community technologies are the main feature of this scenario.
3. Centralization	The main goal is a strong connection to the development of Mexico City and integration into the planned urbanization.	All infrastructure services are centralized as much as possible.

TABLE 5: Examples for technologies with a focus on reuse and recycling.

Resource	Proposed treatment technology
Water	Rain Water Harvesting;
	Constructed wetlands;
	Compact treatment plants (e.g., Biostar);
	(Biological) filters to treat grey wastewater or canal water prior to use.
Energy	Anaerobic digestion of organic waste
Nutrients	EcoSan systems;
	Composting;
	Vermicomposting.

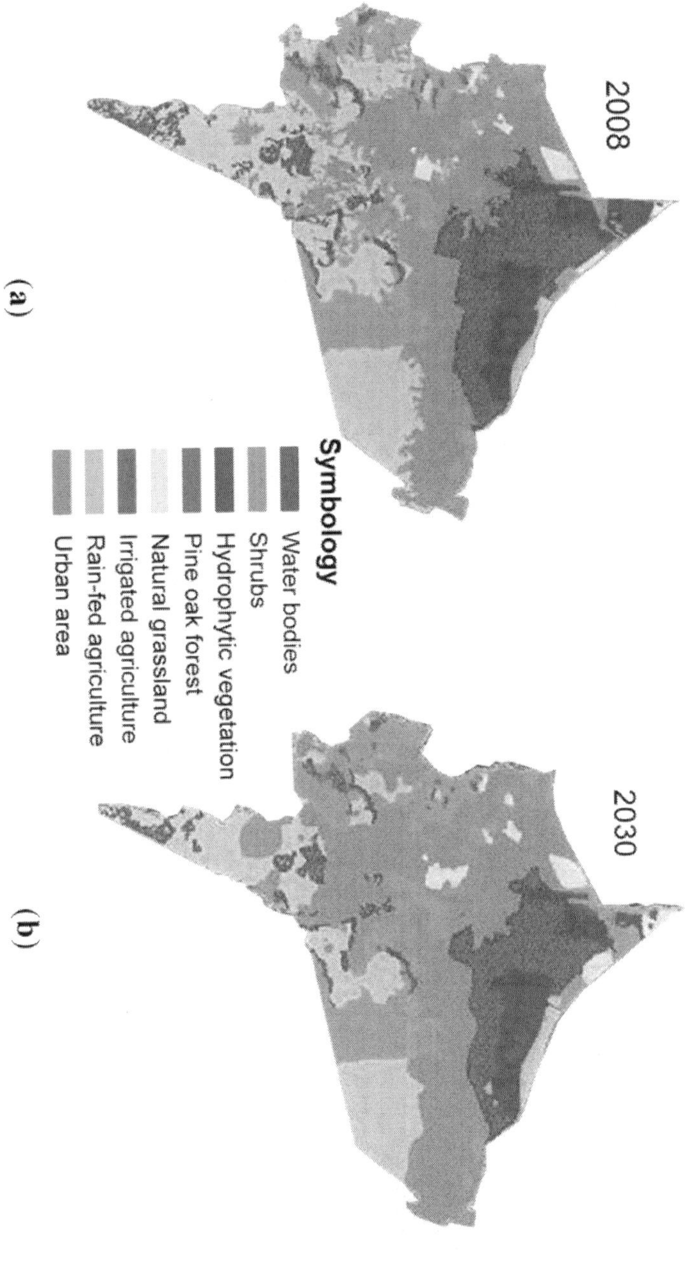

FIGURE 3: (a) Map of Xochimilco 2008; and (b) trend analysis for 2030 (Adapted from [30]).

2.4.3 RESULTS—TECHNICAL FEASIBILITY STUDY

2.4.3.1 WATER SUPPLY

Currently, as rainwater is the only water source not yet exploited in Xochimilco, special attention was paid to the technologies focused on this. The harvesting, storage and treatment (UV treatment, chlorination) of rainwater to supply water for households during the raining season, in combination with gabion dams that are constructed at strategic locations and thereby facilitate the infiltration of collected surface runoff, are two technologies that make optimal local use of the rainwater. The use of rainwater in the households lessens the demand for water from other sources (i.e., the aquifer, surface water bodies) during a certain time of the year, and the infiltration of the remaining rainwater enables the replenishment of the aquifer and hence storage of the water for times when the water demand is higher than the available precipitation. The use of treated wastewater for non-potable domestic and agricultural purposes also lessens the demand for potable water. In general little polluted wastewaters, such as grey wastewater, can be treated by decentralized wastewater treatment systems before reuse without posing any environmental or health risks. Water from canals can also be used for specific non-potable domestic purposes such as cleaning or watering of ornamental plants. The canal water in Xochimilco is too polluted to be used directly for domestic purposes, but can be treated by for instance a constructed wetland or small water filter depending on its quality and the proposed use. Because the main precipitation in Xochimilco falls in six months (May till October), the existing centralized water supply system is expected to remain a very important source for drinking water. The conventional water supply system uses groundwater as its source, which indicates the importance of groundwater replenishment technologies. The rehabilitation and expansion of the water supply system by, for instance, legalizing the illegal connections to the water supply lines and ensuring that the existing (illegal) water pipes are properly constructed and maintained and that taps are not left to run unattended, hence remains an important action. The following example shows how the demand for water from the centralized water supply system can be lessened through the implementation of on-site rainwater harvesting technologies with UV-treatment prior to use or consumption.

(1) Description of Current Situation

Although workshop participants have, amongst others, proposed the use of rainwater harvesting technologies (RWH) (see Section 4.1 and Table 5), at the moment these are not applied in the case study area. This can be attributed to the lack of capacity needed to construct and operate such a system, unwillingness to implement a new technology and/or the poor state of several houses. Houses in the peri-urban areas are built with a variety of materials, including cement, ceramic tiles and corrugated iron sheets, asbestos sheets, cardboard and wood. In general, inhabitants of the peri-urban areas do not pre-treat their water before consuming unless boiling. The application of tUVo's (a UV-treatment technology developed in México) and chlorination of water prior to use is currently not common.

(2) Potential Application in Case Study Area

The harvesting, storage and utilization of rainwater at domestic level is an alternative to avoid the overexploitation of the underground aquifers and the surface water sources in the peri-urban areas of the México City. This will be possible in the raining season and part of the dry season (December). The average annual precipitation in Mexico City is 807 mm [29], with the majority falling in June till October. The collected rainwater will be stored in storage tanks before use. The capacity of these storage tanks will depend on the water demand as well as if there are also connections to the centralized water supply system. As the average weekly household water demand is 0.8 m^3 (with each household consisting of 4 individuals and based on households with flush toilets as well as pit latrines) a total water amount of 3.5 m^3 is needed per month. In general, the roof surfaces of houses in the peri-urban areas of Xochimilco are estimated to be around 36 m^2 [31]. Assuming a 70% collection efficiency (losses and diversion of the first flush) 4.3 m^3 of rainwater can be harvested in the month with the highest monthly precipitation (July; 172 mm) and 0.2 m^3 in the month with the lowest precipitation (December; 7 mm). The highest monthly RW supply is almost equal to the monthly water demand, which means that the storage tank can be kept small as water will be used the same month it is collected. The highest daily maximum precipitation is around 63 mm, yielding a collected amount of 1.6 m^3. By installing a storage tank with the minimum 1.5 m^3 for each household, it is ensured that all collectable rainwater can be utilized optimally.

The installation of RWH technologies can result in the improvement of the living conditions of the inhabitants due to the improved roofs and supporting structures indirectly, and water saving directly. This will also affect the environment indirectly, as less water will be extracted from the aquifer or canals.

The installation of on-site RWH technologies will require the adaptation or improvement of houses. The main adaptations required will be new roofs and supporting structures if these are not suitable for rain water harvesting. In addition, gutters, transportation pipes and a storage tank will need to be installed. Installing the storage tank on a structure (but below the level of the roof) the water can be collected with pales or transported into the house with a pipeline. The water can be treated in the house with filtration, and/or a UV-light (a tUVo, requiring energy) or chlorination.

2.4.3.2 WASTEWATER TREATMENT

The use of EcoSan toilets could improve the sanitation conditions of the households and do not require water, as is the case with flush toilets, thereby lessening the overall household water demand. This is an advantage due to the current aquifer depletion taking place. Lastly, the advantage of EcoSan is that the feces and urine can be co-composted with other organic wastes or the effluents can be applied directly on the fields (chinampas or greenhouses), thereby containing resources in the area. Discharged wastewater that is not very polluted, such as grey wastewater, can be treated by constructed wetlands that can be constructed in the yards of the houses.

Another possibility is to collect the wastewater and treat it with a decentralized wastewater treatment system such as a constructed wetland. These can be constructed in gullies or drains where wastewater is collected or discharged to. Another option is to implement a more complex technology such as a Biostar, which is a biological wastewater treatment, designed by the Instituto Mexicano de Tecnología del Agua (IMTA) in Cuernavaca, México, or an Rotating Biological Contactor (RBC). However, these will need a (small) sewage network and perhaps pumps to collect and transport the wastewater besides the need for skilled operators as well as an energy source to function and perform adequately.

In a more centralized scenario the existing sewage system and wastewater treatment plant can be extended. However, the multiple unplanned settlements not only make it difficult to extend the existing sewage system in order to serve all households (as the capacity of the wastewater treatment plants might need to be increased and a new sewage network will need to be constructed amidst existing buildings and chinampas, thereby severely limiting the design possibilities), but there are also multiple buildings constructed over the existing sewage system which makes rehabilitation and extension works impossible to perform without demolishing and reconstructing the buildings. Hence the expansion of the centralized system might be more difficult to realize than the construction of decentralized measures. On the other hand, the centralized system is a technology preferred by the local inhabitants, meaning that the implementation of decentralized sanitation options might be more difficult due to the lower social acceptance. This, however, still needs to be evaluated. An example of how the implementation of a decentralized sanitation option such as EcoSan can result in resource recovery and water conservation is presented as follows:

(1) Description of Current Situation

The sanitation situation in the peri-urban areas of Xochimilco differs per location. A number of houses, generally located near the urban areas and that are connected to the water supply system or store water in tanks, make use of flush toilets. These discharge their water into the canals or septic tanks, which are often poorly constructed or just a hole in the ground. Other, more remote, households make use of pour flush toilets or pit latrines that are often poorly constructed or defecate and urinate directly in the environment. Gastronomical and skin problems and diarrheal diseases are common in these areas [32].

(2) Potential Application in Case Study Area

By constructing EcoSan toilets initially at households that currently do not have sanitation facilities or make use of pit latrines access to proper sanitation facilities is improved. The combination with a biofiltro, which is the local name for a small constructed wetland designed to treat grey wastewater at household level, will ensure that all wastewater originating in a household is adequately treated. As it is generally the poor who do not have proper access, their livelihoods and the overall public health is impacted directly and improved. In addition, the recovery and reuse of nutrients can not only lead to less demand for artificial fertilizer (or if no fertilizer was used, higher yields), but it can also improve the soil conditions and thereby the sustainability of the land.

An EcoSan toilet can be constructed in the yard of a household, or as an extension of a house, where in general the dimensions of a toilet or pit latrine (1–1.5 m^2) can be maintained. Care should be taken to design the EcoSan toilet in such a way that it uses energy from the sun to dry the collected feces. Feces and urine will be stored in two separate containers prior to their use or the co-composting. The feces and urine can be used in local gardens, greenhouses or chinampas in the form of compost. A biofiltro, with a surface area of 2 m^2 to treat 100 L of grey water (total dissolved oxygen demand of 555 mg/L) per day [11], can also be constructed in the yard of a house. In certain areas this is already common, and here they are constructed in the form of flowerbeds. Cases where ornamental flowers (*Zantedeschia aethiopica* and *Cannaflaccida*) are used, with good treatment results, are also known [33].

2.4.3.3 SOLID WASTE MANAGEMENT AND AGRICULTURE

Organic waste management and agriculture are treated in one section as the outputs of the organic waste treatment often can be used in agriculture.

Waste separation practices are considered a key practice for solid waste management. By separating the different wastes, some can be reused or recycled, while others can be used as an input for composting in order to retain nutrients in the area. If the waste is collected by a centralized organization, the inhabitants will need to store their wastes in separate closable bins till the wastes are collected. The wastes can then be recycled, composted or disposed of in a location specifically adapted to the purpose. For instance, a centralized composting plant, using vermicomposting, could treat organic wastes from an area and in turn supply the farmers of that area with a high-quality compost, soil improver or organic fertilizer. Depending on the local demand for energy, it is also possible to digest the organic wastes with manure coming from livestock to produce biogas. This can be done on household level or in a centralized way. The effluent of the biodigester can be used for fertigation and composting purposes, or organic fertilizer can be produced.

On-site composting and vermicomposting can be implemented where there is no, or very poor, waste collection facilities as the produced compost can be handled safely. These practices can also be done by families with gardens so that they can reduce their fertilizer demands. Although vermicomposting is more efficient and can eventually produce valuable side-products such as worms and proteins, it also requires more economical input (worms—i.e., *Eisenia fetida*—need to be bought, a structure needs to be constructed to contain the compost) to start-up, and thus by studying both technologies local inhabitants can choose which composting type they prefer and can afford. The location of the (vermi-)composting can be in the vicinity of the house, to ease organic waste disposal, but at a large enough distance to prevent the occurrence of noxious odors or pests. The following is the presentation and discussion of vermicomposting as an example of on-site organic waste management:

(1) Description of Current Situation

The separation and recycling of solid waste is an uncommon practice in the peri-urban areas of Xochimilco. Although a number of inhabitants have access to the centralized collection service, who collect the solid waste with trucks, there are also households who do not make use of these facilities, are not or irregularly visited by the trucks or simply do not make use of the service. Hence solid waste is also burnt or discarded to the streets and environment, resulting in pollution.

(2) Potential Application in Case Study Area

The local separation of solid waste and local vermicomposting of organic solid waste will result in an overall decrease of the amount of waste to be collected. This, in turn, will lower the strain on the municipal waste collection service so that they can extend their services. Another advantage is that awareness about waste production and treatment is created with the inhabitants, making them more aware of their actions and hence less inclined to pollute the environment. Lastly, valuable nutrients are conserved and can be used in the chinampas and gardens as natural fertilizer and soil conditioners.

Vermicomposting is technologically relatively simple to conduct. It will, however, require a behavior change as solid waste will need to be separated. Also, the organic solid waste will be composted near the house, in the yard, which is something that not all households are eager to do, especially as the compost pile might need to be over-turned for active aeration. If it is done, one will need to construct a box (usually 1 m^3). The composting pile will need to have a relative humidity of 70%, and an average temperature of 21 degrees Celsius for the reproduction of the worms. However, during the composting process temperatures can exceed 60 degrees Celsius, which will result in a reduction of pathogens. During (vermi-)composting in general a mass reduction of 10%–30% can be expected [34,35].

2.4.4 RESULTS—RESOURCES FLOW ANALYSIS

Based on the three concept scenarios identified during the workshop and the technologies allocated to each concept scenario resources flows were identified and calculated.

The goal of the concept scenario "Local Identity" is the conservation of local values which is related to the cultivation of chinampas and the prevention of external influences. In this concept scenario individual technical solution are preferred over centralized ones to become more independent from Mexico City.

By implementing multiple on-site technologies waste streams can be collected and treated separately (Table 6). In this concept scenario various resources flows between the sectors will take place: nutrients in the form of urine and decomposed feces as well as treated wastewater and compost will be used in agriculture. The digestion of organic waste produces biogas which is an additional local energy source. This will not only enable the recovery and reuse of water, nutrients and energy, but also lessen the demand for centralized services and hence create more independence for the users.

TABLE 6: Overview of technical systems for concept scenario1—Concept of natural resources flows.

Concept scenario 1—Local identity			
Water supply	Wastewater	Agriculture	Solid waste
- RWH (on-site and groundwater recharge	- Separation black water/grey water	- Conservation of chinampas	- Separation organic and inorganic waste
- Aquifer recharge through rainwater capture	- Suitable treatment for black waters (on-site treatment)	- Use of compost - Local gardens (to support subsistence with vegetables)	- On-site (vermi) composting
- Reuse of treated wastewater for non-drinking purposes	- Treatment of grey water for reuse	- Use of treated wastewater	- Anaerobic digestion of organic waste

The objective of the concept scenario "Economic and community development" is economic development with a strong focus on agriculture. In the mountainous areas where no agriculture is practiced, there is a focus on community development. In this scenario (Table 7) the main resources flows emanate from the sanitation systems that allow the reuse of treated wastewater in the chinampas or in some other areas to improve the agricultural production and to have better tourist services. The compost of the decentralized composting facility is used in local agriculture. In the mountainous area, community technologies are the main feature of this scenario.

TABLE 7: Overview of technical systems for concept scenario 2—Concept of natural resources flows.

Concept scenario 2—Economic and community development			
Water supply	Wastewater	Agriculture	Solid waste
- Communal RWH - Use of treated water from channels for non-drinking purposes	- Community scale wastewater treatment plant (reuse in agriculture or for nondomestic purposes)	- Reuse of water and nutrients for chinampas for agricultural production - Manure and other residues to composting	- Separation of organic waste - Decentralized composting together with sediments from chinampas and other agricultural residues - Collection of inorganic waste, recycling and disposal

The advantage of implementing technologies on communal scale is that larger quantities of water and nutrients can be recovered at one location. Hence it is possible to reuse these resources as inputs for agriculture. In addition, side benefits, such as a business in composting, can be established. However, it is more of a challenge to separate all waste streams from the source as certain wastes have specific challenges. Urine, for example, does not allow for transportation over long distances through pipes as it can result in clogging due to precipitation of minerals (primarily Calcium and Magnesium Phosphates) [36].

In the concept scenario "Centralization" the main goal is a strong connection to the development of Mexico City and integration into the planned urbanization. All infrastructure services are centralized as much as possible.

In a centralized scenario it is much more difficult to recover resources from wastewater and solid waste (as shown in Table 8). The only resources flow between the sectors is through the organic fraction, which can be composted in a centralized location before being solid back to the farmers. The treated wastewater will be discharged to the environment.

Examples of resources flows from earlier discussed technologies are as follows: Through the implementation of an on-site RWH system with post-treatment the inhabitants will be less depended on the centralized water supply system for their water supply. If all annual precipitation (807 mm) is collected with a 70% efficiency on the before mentioned roof surface (36 m^2), this can result in a capturing of 20.3 m^3, which accounts for 49% of the annual water demand of a peri-urban household not connected to the centralized water supply system (41.6 m^3/year) [32].

TABLE 8: Overview of technical systems for concept scenario 3—Concept of natural resources flows.

Concept scenario 3—Centralisation			
Water supply	Wastewater	Agriculture	Solid waste
- Connection to centralized water supply	- Connection to centralized sewer system and treatment	- Use of compost in local gardens/agriculture	- Separation of organic waste - Centralized composting of organic waste

Examples of resources flows from earlier discussed technologies are as follows: Through the implementation of an on-site RWH system with post-treatment the inhabitants will be less depended on the centralized water supply system for their water supply. If all annual precipitation (807 mm) is collected with a 70% efficiency on the before mentioned roof surface (36 m^2), this can result in a capturing of 20.3 m^3, which accounts for 49% of the annual water demand of a peri-urban household not connected to the centralized water supply system (41.6 m^3/year) [32].

The implementation of an EcoSan and biofiltros for the treatment of feces, urine and grey water at household level will enable the recovery of N, P and Potassium (K) as well as water. Based on data for one household the recovered amounts can be 5.7 kg N/year, 1.5 kg P/year and 2.4 kg K/year. Based on N-requirements, this is enough to cultivate 750 m^2 of hybrid maize capable producing 4.5 t/ha (application load of 80 tons/ha). However, if P-availability is limiting, the cultivatable area is 500 m^2. For K this is 300 m^2 [37,38]. This can deviate, depending on local conditions.

If households that currently make use of flush toilets (using an average of 7 L per flush) also construct EcoSan toilets, water can be saved. With an average of 4 inhabitants per household and 5 flushes per person per day (one flush per visit), an annual 51 m^3 can be saved per household. The use of a biofiltro will ensure that little polluted wastewater can be reused in a safe way for, for instance, irrigation of crops or non-potable domestic purposes such as laundry or cleaning. The annual production of grey wastewater for a household is 112 L per household per day, which will result in an annual amount of 31 m^3 of treated wastewater, assuming a 75% recovery and treatment efficiency.

TABLE 9: Summary of examples of on-site technologies and potential resources recovered.

Technology	Size	Potential resource recovery per household per year	Benefit
On-site rainwater harvesting with posttreatment	Roof: 36 m^2 Storage tank: 1.5–2.0 m^3	Water: 20.3 m^3 (70% efficiency)	49% reduction in water demand
EcoSan and biofiltro	EcoSan: 1.5 m^2 Biofiltro: 2 m^2	N: 5.7 kg P: 1.5 kg K: 2.4 kg Flush water: 51 m^3 Treated grey water: 31 m^3 (75% efficiency)	300–750 m^2 of maize, producing an estimated 4.5 tons/ha Redundant flush water: 51 m Recovered grey water accounts for 60% of the water demand
(Vermi-)composting	Composting structure: 1m^3	N: 6 kg P: 1 kg OM: 200 kg Worms: 10 kg from 1 kg/year	300–750 m^2 of maize, producing an estimated 4.5 ton/ha

The annual amount of organic waste produced per household in Xochimilco is approximately 1000 Kg. Assuming a 20% mass reduction this results in an annual production of 800 kg of compost from the organic waste of one household. The composition of compost generated will differ per location as it depends upon the organic waste that is composted, the quality of the composting facility (aeration, temperature regulation) as how the compost is maintained (moisture content and temperature). Taking the general values shown in Table 2 the amount of nutrients recovered annually by one household is 6 kg N and 1 kg P. Compost is also a source for organic matter (OM), which can be used as soil conditioner. The annual production of OM is calculated to be 200 kg OM. Based on N-requirements, this is enough to cultivate 750 m^2 of hybrid maize capable producing 4.5 ton/ha (application load of 80 ton/ha). However, if P-availability is limiting, the cultivatable area is approximately 300 m^2 [38]. Apart from the nutrients recovered through the composting process, vermicomposting also allows for the cultivation of worms that can be sold to start-up other vermicomposting sites or as protein sources for animal feeds. In one year time the amount of worms can increase by a ten-fold, and one kilogram of worms (*Eisenia fetida*) will sell for about € 15 to €60, depending on the demand and quality [39].

The different technologies as well as different resources recovered are summarized in Table 9.

2.5 DISCUSSION

The participatory planning approach combining scenario development (using a Future Workshop) with participatory planning has shown to be a useful approach to raise awareness and interest of users and stakeholders to the issues at stake. Complex technical issues could successfully be discussed with a wide range of local users and stakeholders.

When approaching the water supply, sanitation, wastewater treatment and solid waste management from the resource conservation and reuse point of view a different approach to the water chain and a different interpretation of the characteristics of the water and waste is required. By following this reverse water chain approach the potential of pollutants can be seen as valuable resources, both in terms of physical resources (nutrients or water), but also as economic resources. The brief descriptions of three different decentralized technologies show that the implementation of all three technologies can result in an improvement of the living conditions of the inhabitants of peri-urban areas while there are also economic savings as less water or fertilizers will have to be bought and one will need to pay less to a centralized body for the treatment of one's water or wastes. In addition, the individual technologies presented require little space (1–3 m^2) and hence generally fit within the gardens of the local houses, even if all three are constructed at one house although this can differ per case, and can be constructed with materials available locally in México. The fact that there is no need for large excavation for pipelines or external energy sources for pumps and regulators is another technological advantage of the decentralized technologies because it increases its versatility by making it applicable for areas not easily accessible and lessens construction costs of large sewage systems.

The implementation of decentralized technologies does not by definition mean that this is on a household, or on-site, scale. If the local context allows for this, technologies could also be implemented on communal level and thereby create small economies of scale and density with all benefits (and constraints) of this. Another advantage of decentralized technologies is that different technologies can be combined on different scales to achieve different objectives. The example presented earlier considered three specific technologies, but the research also showed that, for instance, the implementation of constructed wetlands on communal level can provide water suitable for supplementary irrigation while the digestion of organic wastes at household level provides local energy for cooking, lighting or heating. The sludge can be used as a natural fertilizer and soil conditioner, and hence an impulse to the local economy through agricultural stimulation is created.

However, as mentioned before, the implementation of the decentralized technologies and reuse of resources will require the acceptance of the technologies by society as well as a change in perception and behavior from society. In addition, it might initially cost them money as they will need to construct the different technologies. It is not self-evident that the use of collected rainwater and treatment prior to consumption (chlorine or UV-light if there is an energy source), as well as the use of treated wastewater for non-potable purposes is accepted and implemented readily by all inhabitants. Furthermore, the use of an EcoSan toilet could be perceived as less comfortable, easy or less hygienic than a conventional flush toilet. As a decentralized technology will require operation and maintenance by the inhabitant, one is directly responsible for the wastes that one produces. Hence decentralized technologies demand an active involvement of the users, for instance in the form of cleaning the roof and rainwater storage tanks, handling the (dried) feces and urine, separating wastes and periodically over-turning the compost pile. This active involvement often has a low social acceptance if there is an easier option, such as a centralized, conventional, system, available. Thus, the society plays an important role in implementing decentralized technologies as their cultural perceptions, values and wishes will need to be taken into account while designing the technology or combination of technologies, which is not self-evident [40,41].

Another obstacle for the implementation of decentralized technologies and the reuse of resources could be institutional constraints or legislation. Legislation could restrict the use of human excreta as a natural fertilizer or pose restrictions on the quality of the effluent if it is to be discharged to a natural water body. In addition, the implementation of centralized technologies can be used by authorities to gain control over resources flows to an area and hence the ability to influence its population. In the case of decentralized technologies this is much less of a possibility, and hence authorities could be less inclined to promote the approach proposed here.

The last obstacle discussed in this paper is the threat of the technologies to the public health. Although the technologies are envisaged to be implemented in peri-urban areas where there is inadequate access to safe water supply, wastewater treatment and solid waste management services, and thus improve the living conditions of the inhabitants, improper construction, operation or maintenance can result in serious health risks. This is especially the case for EcoSan, which if not constructed, operated or maintained properly can become hygienically unsafe and result in a source of pathogens or vectors, thereby increasing the risk of infections.

2.6 CONCLUSIONS

As the research results have shown, the implementation and combination of decentralized technologies allows for context-specific solutions to local water supply, sanitation or solid waste dilemmas that are often relevant in peri-urban areas. Moreover, it provides the opportunity to lessen environmental pollution and recover valuable resources to sustainably improve livelihoods and the environment. However, in order to achieve this, participation of users and stakeholders is crucial. As the experiences of this project have shown, stakeholders still often favor traditional centralized solutions and, in order to facilitate the implementation of decentralized solutions, an open and transparent planning process that highlights advantages and possible disadvantages of decentralized over centralized solutions needs to be pursued. At the very beginning of such a planning process a clear vision of the local stakeholders is necessary establishing what the existing problems are and how they could be solved. A participatory planning process linked to the identification of development scenarios (using Future Workshops) as tested in this study can help to develop technical concept scenarios that show all options for problem solving. As a next step, these options need to be carefully assessed not only from an environmental, but also from an economic and social, perspective.

REFERENCES

1. Coles, J.W.; Hesterly, W.S. Transaction costs, quality, and economies of scale: Examining contracting choices in the hospital industry. J. Corp. Financ. 1998, 4, 321–345.
2. Graham, D.J.; Couto, A.; Adeney, W.E.; Glaister, S. Economies of scale and density in urban rail transport: Effects on productivity. Transp. Res. Part E Logist. Transp. Rev. 2003, 39, 443–458.
3. Kraus, M. Economies of scale in networks. J. Urban Econ. 2008, 64, 171–177.
4. Peter-Varbanets, M.; Zurbrügg, C.; Swartz, C.; Pronk, W. Decentralized systems for potable water and the potential of membrane technology. Water Res. 2009, 43, 245–265.
5. Körner, I.; Saborit-Sánchez, I.; Aguilera-Corrales, Y. Proposal for the integration of decentralized composting of the organic fraction of municipal solid waste into the waste management system of Cuba. Waste Manag. 2008, 28, 64–72.
6. Mankad, A.; Tapsuwan, S. Review of socio-economic drivers of community acceptance and adoption of decentralised water systems. J. Environ. Manag. 2011, 92, 380–391.
7. Massoud, M.A.; Tarhini, A.; Nasr, J.A. Decentralized approaches to wastewater treatment and management: Applicability in developing countries. J. Environ. Manag. 2009, 90, 652–659.
8. Zurbrügg, C.; Drescher, S.; Rytz, I.; Maqsood Sinha, A.H.; Enayetullah, I. Decentralised composting in Bangladesh, a win-win situation for all stakeholders. Resour. Conserv. Recycl. 2005, 43, 281–292.
9. Huibers, F.P.; van Lier, J.B. Use of wastewater in agriculture: The water chain approach. Irrig. Drain. 2005, 54, 3–9.
10. Van Lier, J.B.; Huibers, F.P. From unplanned to planned agricultural use: Making an asset out of wastewater. Irrig. Drain. Syst. 2010, 24, 143–152.
11. Nanninga, T.A. Helophyte Filters, Sense of Non-sense? A Study on Experiences with Helophyte Filters Treating Grey Wastewater in The Netherlands. Master's Thesis, Wageningen University, Wageningen, The Netherlands, July 2011.
12. Merlín-Uribe, Y.; Contreras-Hernández, A.; Astier-Calderón, M.; Jensen, O.P.; Zaragoza, R.; Zambrano, L. Urban expansion into a protected natural area in Mexico City: Alternative management scenarios. J. Environ. Plan. Manag. 2012, in press.

13. Wigle, J. The "Xochimilco" model for managing irregular settlements in conservation land in Mexico City. Cities 2010, 27, 337–347.
14. Genovevo, J.P.E. Chinampas: Entre apantles y acalotes (in Spanish). In Los Pueblos Originarios de la Ciudad de México: Atlas Etnografico; Mora, T., Ed.; Gobierno del distrito Federal and Intitutio Nacional de Antropologia e Historia: Mexico City, México, 2008; pp. 97–101.
15. Martínez-Ruiz, J.L. Manual de Construcción de Chinampas (in Spanish); Instituto Mexicano de Tecnología del Agua (IMTA): Cuernavaca, México, 2004.
16. Jimenéz-Osorina, J.I. Past, present and future of the chinampas. In Maya Sustainability Report; Dry Lands Research Institute, University of California: Riverside, CA, USA, 1990.
17. González-Pozo, A. Las Chinampas de Xochimilco al despuntar el siglo XXI: Inicio de su Catalogación (in Spanish); Universidad Autónoma Metropolitana: Mexico City, México, 2010.
18. The Ramsar Convention on Wetlands Homepage. The Ramsar list of Wetlands of International Importance. Available online: http://www.ramsar.org/cda/en/ramsar-documents-list/main/ramsar/1- 31-218_4000_0__ (accessed on 27 June 2012).
19. Martínez-Ruiz, J.L. Origen, grandeza y decadencia de la chinampa: Crónica de un colapso socioambiental anunciado (in Spanish). In Proceedings of the 12th National Irrigation Congress (Congreso Nacional de Irrigación), Zacatecas, México, 13–15 August 2003; ANEI 2003 M6 T14; pp. 107–116.
20. Programa de Desarrollo Urbano de la Delegación Xochimilco (in Spanish); Secretaria de Desarrollo Urbano y Vivienda: Mexico City, México, 2005.
21. Bernstein, J. Toolkit for Social Assessment and Public Participation in Municipal Solid Waste Management; Urban Environmental Thematic Group, World Bank: Washington, DC, USA, 2004.
22. Bromell, D.; Hyland, M. Social Inclusion and Participation: A Guide for Policy and Planning; Ministry of Social Development: Wellington, New Zealand, 2007.
23. FAO. Participatory Communication: A Key to Rural Learning Systems; Food and Agricultural Organization: Rome, Italy, 2003.
24. Jungk, R.; Müller, N. Future Workshops: How to Create Desirable Futures; Institute of Social Inventions: London, UK, 1987.
25. Qué Sabemos de Orina Humana Como Fertilizante—What we know about human urine as a fertiliser (in Spanish). Available online: http://www.sarar-t.org (accessed on 11 June 2012).
26. Gener Cada Mexicano Un Kilogram De Baura El Día. México: Noticias 8 May 2011 (In Spanish). Available online: http://noticias.universia.net.mx/en-portada/noticia/2011/08/05/854479/genera-cadamexicano-kilogramo-basura-dia.html (accessed on 18 June 2012).
27. Enviros RIS. Residential Waste Materials Density Study; No. WDO OPT/ORG-R2-02; Waste Diversion Organization: Toronto, Canada, 2001. Available online: http://www.wdo.ca/files/domain4116/Access%20OPT_ORG%20R2-02.pdf (accessed on 18 September 2012).
28. Jönsson, H.; Baky, A.; Jeppsson, U.; Hellström, D.; Kärrman, E. Composition of Urine, Faeces, Greywater and Biowaste for Utilisation in the Urware Model; Urban Water, Chalmers University of Technology: Göteborg, Sweden, 2005.

29. Normales Climatológicos 1971–2000, Estación San Gregorio Atlapulco—Climatological normals 1971–2000, San Gregorio Atlapulco Station (in Spanish). Available online: http://smn.cna.gob.mx (accessed on 27 June 2012).
30. Instituto Mexicano de Tecnología del Agua (IMTA). Personal Communication with Geographer during a workshop in San Gregorio Atlapulco (Xochimilco) for the Project Vivace; Instituto Mexicano de Tecnología del Agua (IMTA): Cuernavaca, México. Personal communication, 2010.
31. Instituto Mexicano de Tecnología del Agua (IMTA). Personal Communication with Expert and Survey Conductor; Instituto Mexicano de Tecnología del Agua (IMTA): Cuernavaca, México. Personal communication, 2012.
32. Instituto Mexicano de Tecnología del Agua (IMTA). Detailed Household Surveys Conducted in La Conchita and San Martín (Xochimilco) for the Project Vivace; Instituto Mexicano de Tecnología del Agua (IMTA): Cuernavaca, México, 2011.
33. Belmont, M.A.; Cantellano, E.; Thompson, S.; Williamson, M.; Sánchez, A.; Metcalfe, C.D. Treatment of domestic wastewater in a pilot-scale natural treatment system in central México. Ecol. Eng. 2004, 23, 299–311.
34. Breitenbeck, G.A.; Schellinger, D. Calculating the reduction and material mass and volume during composting. Compost Sci. Util. 2004, 12, 365–371.
35. Yue, B.; Chen, T.; Gao, D.; Zheng, G.; Liu, B.; Lee, D. Pile settlement and volume reduction measurement during forced-aeration static composting. Bioresour. Technol. 2008, 99, 7450–7457.
36. Tilley, E.; Lüthi, C.; Morel, A.; Zurbrügg, C.; Schertenleib, R. Compendium of Sanitation Systems and Technologies; Swiss Federal Institute of Aquatic Science and Technology (Eawag): Dübendorf, Switzerland, 2008.
37. Dierolf, T.; Fairhurst, T.; Mutert, E. Soil Fertility Kit: A Toolkit for Acid, Upland Soil Fertility Management in Southeast Asia; Singapore Potash & Phosphate Institute (PPI): Singapore, 2001.
38. Richert, A.; Gensch, R.; Jönson, H.; Stenström, Dagerskog, L. Practical Guidance on the Use of Urine in Crop Production; EcoSanRes Series 2010-1; Stockholm Environmental Institute (SEI): Stockholm, Sweden, 2010.
39. Gómez, L.M. Manual De Lombricomposta (in Spanish); Secretaría de Educación Publica (SEP) and Colegio de Estudios Cientificos y Tecnológicos del Estado de Chiapas: Tuxtla Gutiérrez, México, 2008.
40. Martínez-Ruiz, J.L.; Murillo, D. The problem of the social adoption of appropriate technologies. In Revista Plurimondi: An International Forum for Research and Debate on Human Settlements; Edizioni Dedalo: Bari, Italy, 2010; Volume 4, pp. 125–145.
41. Starkl, M.; Bisschops, I.; Norström, A.; Purnomo, A.; Rumiati, A. Integrated assessment of the feasibility of community based sanitation options: A case study from East Java, Indonesia. Water Pract. Technol. 2010, 5, 1–12.

PART II

SOIL AND LAND

CHAPTER 3

EVALUATION OF POST-HARVEST ORGANIC CARBON AMENDMENTS AS A STRATEGY TO MINIMIZE NITROGEN LOSSES IN COLE CROP PRODUCTION

KATELYN A. CONGREVES, RICHARD J. VYN, AND LAURA L. VAN EERD

3.1 INTRODUCTION

Nitrogen fertilizer applications are frequently used to enhance vegetable crop production, and much research has been done to reduce N losses during the vegetable crop growing season [1,2]. However, after harvest, a large quantity of N can remain in vegetable crop residues [2–4], which readily mineralizes [3] and may be susceptible to post-harvest losses. There is need to minimize N losses during the post-harvest season when the risk of losses is much greater due to the annual water budget in Ontario [5] and elsewhere. Losses include nitrate (NO_3^--N) leaching and denitrification, which can have negative environmental consequences such as groundwater and atmosphere contamination. Thus, the development of

This chapter was originally published under the Creative Commons Attribution License. Congreves KA, Vyn RJ, and Van Eerd LL. Evaluation of Post-Harvest Organic Carbon Amendments as a Strategy to Minimize Nitrogen Losses in Cole Crop Production. Agronomy *3 (2013); 181-199. doi:10.3390/agronomy3010181.*

more sustainable agricultural practices, which are focused on soil N management after vegetable harvest, is necessary.

Cole crop (*Brassica* vegetables), in particular, produce optimal yields with high N applications, ranging from 270 to 550 kg N ha^{-1} for broccoli (*Brassica olecerea* var *italica* L.) [4,6–9]. Cole crop residues may leave ≈100 to 330 kg N ha^{-1} in the field at harvest [2,3], and post-harvest mineral N losses are more related to crop residue N rather than N fertilizer remaining in the soil [3]. Considering that 35 to 60% of broccoli crop residue N has been found to mineralize in controlled incubation studies [10,11], and that the crop residue may contain up to 330 kg N ha^{-1} [1], then up to 198 kg N ha^{-1} would be mineralized in the field after harvest from broccoli crop residue [11]. Thus, cole crop residue poses a significant risk for N losses due to the large quantity of mineralizable N in the post-harvest season.

Amending the soil with organic C material has the potential to reduce soil mineral N (SMN) concentrations through N immobilization [11,12]. By redirecting organic C materials from waste streams, a new sustainable method of utilizing these materials could be developed. Some high C materials, which are expected to be readily available for vegetable producers in Ontario, are wheat straw, yard waste, or used cooking oil. For example, winter wheat typically has 1.4 Mg ha^{-1} of straw residue [13]. Up to 326,000 ha^{-1} of winter wheat was harvested in Ontario in 2010 [14]. Also, 400,000 Mg yr^{-1} of leaf and yard waste has been estimated from Ontario residential collection [15], and approximately 450,000 Mg of oily food waste is produced annually in Ontario [16,17]. Considering that the estimated total production of broccoli, cabbage, and cauliflower in Ontario was on 4147 and 4247 ha^{-1} in 2010 and 2011 [18], respectively, there is potential for incorporating the organic C waste materials as amendments for N management after cole crop production.

Research has demonstrated that SMN concentrations may be reduced via N immobilization by the applications of wheat straw [19–21], yard waste [22–24], and oily food waste [16,25,26]. In the aforementioned studies, immobilized N was derived from fertilizers or indigenous SMN. The synchrony of cole crop and organic C material decomposition is crucial for the immobilization of N derived from crop residue. In an incubation study, N derived from broccoli crop residue was immobilized by the addition of wheat straw, yard waste, and used cooking oil [11]. The rapid decomposition and synchrony of the crop residue N mineralization rate relative to the organic C amendment (OCA) decomposition suggested promise for field application [11].

Even though a reduced potential for N losses may result if the period of N immobilization coincides with periods of high risk for N losses in the field, it is necessary to assess potential effects of OCA on the subsequent crop. Early- rather than late-season N fertilizer input is recommended to achieve a desired yield goal for spring wheat (*Triticum durum* L.) production [27]. Therefore, if N immobilization due to the OCA after cole crop harvest is not followed by re-mineralization early enough in the spring wheat growing season, a negative effect on spring wheat yield could result. Conversely, if N mineralization is synchronous with spring wheat N demand, then N use efficiency may be enhanced.

A pattern of N immobilization followed by re-mineralization has been found in previous laboratory [11] and field [26] research after the OCA of used cooking oil. However, other researchers did not observe re-mineralization in the subsequent spring after autumn incorporation of cereal straw or green compost with cauliflower crop residue [12]. Research has found corn, lettuce, or leek production to be unaffected by the previous autumns' application of oily food waste [16], cereal straw or waste compost [12], respectively. Thus, it may be possible to reduce potential N losses during periods of high risk for N losses by applying an OCA, without negatively affecting the subsequent crop.

In addition to the environmental impacts of N management practices, consideration must be given to the economic impacts. Some studies have found a trade-off to exist between environmental and economic benefits because practices that reduced N losses did not have favorable economic outcomes in vegetable production [28,29]. Other studies have found an overlap for the optimal environmental and economic outcomes [30,31]. Thus, ambiguity exists in the literature with respect to the economic impacts associated with management practices for reducing N losses. Regardless, the economic implications of proposing a better management practice to minimize N losses should be evaluated.

Therefore, the objective of this field study was to evaluate the effects of three different OCAs of wheat straw (OCA-straw), yard waste (OCA-yard), and used cooking oil (OCA-oil) on N dynamics, spring wheat production, and profit margins following broccoli harvest, compared to typical grower practices. This research could lead to the development of better N management practices, leading to more sustainable cole crop production by minimizing environmental N contamination.

3.2 MATERIALS AND METHODS

The field sites at Ridgetown Campus in 2009–2010 and 2010–2011 were on a Brookston clay loam (Orthic Humic Gleysol) soil, with textures of loam and sandy-clay-loam, respectively (Table 1). The soil characteristics evaluated (Table 1) included pH (1:1 v/v method), organic matter (modified Walkley Black method), N (KCl extraction and colorimetric analysis method), P (Olsen bicarbonate extraction method), Ca, K, Mg (atomic absorption via ammonium acetate extraction), cation exchange capacity (CEC) (estimated based on ammonium acetate extraction and pH), and soil texture (hydrometer method) [32].

TABLE 1: Initial soil characteristics of the experimental sites prior to broccoli transplanting in 2009 and 2010.

Characteristics (soil sample depth 15 cm)	2009	2010
pH	6.1	5.6
Soil texture	Loam	Sandy Clay Loam
Sand:Silt:Clay (%)	46:28:26	58:18:24
OM (Mg ha^{-1})	55	80
CEC (cmol kg^{-1})	15	23
Bulk density (g·cm^{-1})	1.4	1.4
Pre-plant nutrients (mg·kg^{-1})		
N	5	41
P	34	37
K	108	87
Mg	153	149
Ca	2433	2430

Broccoli (c.v. "Ironman") was grown in two systems: Early- and late-harvest, which was followed by spring wheat production in the subsequent year. The experimental design was a randomized complete block with four replications within the early- and late-broccoli systems. Nitrogen was uniformly hand-applied at the typical grower rate of 342 kg N ha^{-1} as urea, and incorporated by disking before broccoli was mechanically transplanted on May 23 and 17 for the early system and June 23 and 23 for the late system in 2009 and 2010, respectively. Early- and late-broccoli systems were grown in the same field, but considered as separate research trials. Typical management practices were followed for pre-plant fertilization of macronutrients, plant spacing (75 cm between rows and 30 cm between plants), insecticide application (Lambda-Cyhalothrin (13.1%) 19 mL ha^{-1}), and drip irrigation, as required. Temperature and precipitation data were obtained by the on-site weather station (Table 2).

TABLE 2: Monthly total precipitation and mean temperature and the 30-yr mean at Ridgetown, ON, Canada, during 2009–2011.

	Total precipitation (mm)				Mean temperature (°C)			
	2009	2010	2011	30 yr mean	2009	2010	2011	30 yr mean
January	147	227	61	61	−10.1	−5.5	−7.5	−3.7
February	106	98	137	54	−3.8	−4.8	−6.1	−2.4
March	106	67	88	60	1.1	3.4	−0.3	2
April	152	63	134	78	7.8	9.8	6.6	8.3
May	49	114	153	75	13.0	14.4	14.0	14.8
June	65	97	75	83	17.3	19.3	18.8	20.2
July	31	121	70	86	18.5	22.6	23.5	22.5
August	92	19	70	86	19.6	21.2	20.5	21.4
September	36	80	135	93	16.1	16.2	16.5	17.6
October	70	78	79	69	8.6	10.8	10.4	11.2
November	30	92	140	75	6.2	4.2	6.9	4.8
December	138	45	86	67	−2.5	−4.8	1.6	−1.2
Total	1022	1101	1228	887				

The early-broccoli system was harvested on August 4 and August 3, while the late-broccoli system was harvested on August 31 and September 20, in 2009 and 2010, respectively. Heads were hand harvested from the entire trial, and broccoli yield was estimated by recording head counts and weights in a 3 m harvest row from three random plots per replicate. Broccoli leaf, stem, and head samples were collected from a composite of three plants in each of three random plots per replicate for dry matter and N content determination.

Treatments were applied 1 to 2 d after harvest with plots 9 m long by 3 m wide. The control treatment representing the typical grower practice was the incorporation of broccoli crop residue (CR-control), and was established by mechanically mulching then disking the broccoli crop residue. The crop residue removal (CR-removal) treatment was established by hand-removing all the above-ground biomass from the plots prior to disking. An oat (*Avena sativa* L.) cover crop (CC-oat) was established by drilling seeds at a rate of 108 kg ha^{-1} after the broccoli residue was mulched and incorporated. The OCA treatments were established by uniformly hand-applying either wheat straw (OCA-straw, C:N ratio ≈ 65:1), yard waste (OCA-yard, C:N ratio ≈ 56:1), or used cooking oil (OCA-oil, C:N ratio > 1000:1), at a fresh rate of 5 Mg ha^{-1}` onto mulched broccoli crop residue (C:N ratio ≈ 11:1), followed by disking. In addition to samples at broccoli harvest, post-harvest soil samples in autumn were taken for mineral N analysis from a composite of four soil cores per plot at depths of 0–30 and 30–60 cm on August 21, September 18, and October 22 for the early system in 2009, on August 20, September 1, September 22, and October 25 for the early system in 2010, on October 26 for the late system in 2009, and on October 6 and October 25 for the late system in 2010.

In the subsequent growing season, the entire trial area was cultivated and spring wheat seed was drilled at 154 kg ha^{-1}, without N fertilization. A broccoli residue incorporated treatment with pre-plant urea applied at 103 kg N ha^{-1} (CRN-control) was also established as the treatment representing the typical grower practice. Usual management practices were followed for spring wheat production [33]. Grain was mechanically harvested from the center area of 6 by 1 m, and grain and straw were collected to assess the above-ground plant biomass (grain + straw) and grain yield (Mg ha^{-1}), N content (kg N ha^{-1}), and N harvest index (%). Harvest occurred on August 3, 2010 and August 23, 2011. Soil was sampled, as described above, for mineral N analysis from depths of 0–30, 30–60, and 60–90 cm at spring wheat planting and harvest.

3.2.1 NITROGEN MEASUREMENTS

Plant N content (broccoli and spring wheat) was determined by dry combustion method [34] using a LECO CN analyzer (Leco Corporation, St. Joseph, MI, USA), following grinding with a 2 mm diameter mesh screen opening on a Wiley Mill (Thomas Scientific, Swedesboro, NJ, USA). Soil NO_3^--N and NH4 +-N concentrations were quantified by the KCl extraction method [35]. Briefly, 5 g of soil was extracted with 25 mL of 2 M KCl, shaken for 30 min, filtered and analyzed colorimetrically on an AutoAnalyzer3 (SEAL Analytical Inc., Mequon, WI, USA) with high resolution digital colorimeter to quantify NH4 +-N (method G-102) and NO_3^--N (method G-200). Soil mineral N was the sum of NO_3^--N and NH_4^+-N and expressed as kg N ha^{-1} based on soil bulk density.

3.2.2 ECONOMIC ANALYSIS

The economic analysis was conducted through a comparison of spring wheat profit margins for each treatment. Profit margins were calculated by subtracting costs that varied across post-harvest treatments from revenues. Revenues were calculated based on spring wheat yields from each treatment and the average spring wheat price in Ontario between 2010 and 2011, as reported by the Ontario Ministry of Agriculture Food and Rural Affairs [36]. Costs that varied included those associated with mulching residue (all treatments except the CR-removal treatment), residue removal, pre-plant N fertilizer for spring wheat crop, seeding cover crop, and applying the OCA as listed in the OMAFRA's Custom Rate Survey [37]. Costs associated with the OCA included transportation and application as well as baling wheat straw. Some assumptions were made regarding the equipment and the associated rates that would be used to conduct the required activities on a large field scale. For example, application costs were based on the rates for manure spreading, while transportation costs were based on trucking rates for wheat straw, yard waste, used cooking oil, and spring wheat grain [37]. In addition, the CR-removal method costs were based on custom rates for forage harvesting [37]. Other costs which factored into the profit margin calculation included the fertilizer cost, which was based on the average spring price reported in surveys of retail outlets across Ontario between 2009 and 2011 [38] and the oat seed cost, which was based on prices provided by seed retailers in southern Ontario. All other costs were assumed to be constant across treatments; thus, they were not accounted for in the profit margin calculation.

3.2.3 STATISTICAL ANALYSIS

Because the analyses of variance (with fixed effects of treatment, system, year, and respective interactions, and a random effect of block (system)) found significant year x system effects ($P < 0.05$) in datasets for SMN in autumn, spring, and spring wheat harvest, a separate analysis of variance for each year and each early- or late-harvested system was necessary (with a fixed effect of treatment and a random effect of block). Other than the late-broccoli-system in 2009 (which had only one sample date), the autumn SMN repeated measures had no treatment by sample day effect ($P > 0.05$) in each system in each year, thus the average of each repeated measurement was investigated. A multiple means comparison to the CR-control and CRN-control (typical practices) with a Dunnett-Hsu adjustment was applied to each dataset. All significant differences were set at $P < 0.05$.

3.3 RESULTS AND DISCUSSION

3.3.1 BROCCOLI HARVEST

Broccoli yields varied by system and year, but not by system × year. The early-broccoli system had fresh yields of 22.2 and 22.9 Mg ha^{-1} and late-broccoli yielded 25.0 and 36.2 Mg ha^{-1}, in 2009 and 2010, respectively.

Similar yields, 16 to 35 Mg ha^{-1}, have been reported in the literature [2,6–9]. The lower yield in the early-broccoli compared to the late-broccoli system agrees with previous research, which also found lower total and head plant mass in summer harvested broccoli compared to autumn harvest [39], likely due to different environmental conditions during the broccoli growth periods. The crop residue (leaves + stems) fresh weight ranged from 50 to 75 Mg ha^{-1}, similar to broccoli and cauliflower crop residue rates in earlier studies [11,12]. The early-broccoli crop residue contained 202 and 247 kg N ha^{-1}, while the late-system residue had 212 and 207 kg N ha^{-1}, in 2009 and 2010, respectively. Similarly, previous research observed above-ground broccoli N accumulation of 96 to 465 kg N ha^{-1}, with fertilizer N application rates up to 500 kg N ha^{-1} [1,2,4,6,40].

At broccoli harvest, 0–30 cm SMN ranged from 68 to 168 kg N ha^{-1}, yet greater quantities of N (207 to 247 kg N ha^{-1}) were in the broccoli crop residue. Thus, between 265 to 415 kg N ha^{-1} remained in the field after harvest, which was consistent with other research [1,6]. Clearly, strategies that minimize N losses during the post-harvest season would be valuable.

3.3.2 SOIL MINERAL NITROGEN IN AUTUMN

Soil mineral N data were investigated, because both NO_3^--N and NH_4^+-N are susceptible to losses. The treatment effect on soil N was similar for NO_3^--N and SMN concentrations, but the soil NH_4^+-N was largely unaffected by treatment. Thus, the SMN results were presented. The 0–30 and the 30–60 cm depths were analyzed separately for the autumn dataset to assess the downward movement and leaching potential of SMN in autumn. The year variation in autumn SMN between 2009 and 2010 could be a residual N effect attributed to the type of crop grown the year prior to broccoli production, which was corn and soybean, respectively. Additionally, the higher autumn SMN variation in the early- compared to the late-broccoli systems may be due to higher temperatures during the early autumn (August to November) which may have permitted more decomposition and resulted in a greater N mineralization, compared to cooler temperatures after the late-broccoli system (September to November).

In the CR-control, autumn 0–30 cm SMN concentration ranged from 118 to 368 kg N ha^{-1} (Figure 1) and SMN was affected by treatment. After the early-broccoli system and relative to the CR-control, the 0–30 cm SMN concentrations were reduced by the OCA treatments of OCA-straw in 2009 and 2010, OCA-yard in 2010 but not in 2009, and OCA-oil in 2009 and 2010 (Figure 1).

After the late-broccoli system, the only treatment which reduced SMN compared to the CR-control was the OCA-oil in both 2009 and 2010 (Figure 1). The CC-oat treatment did not have different 0–30 cm SMN concentrations compared to the CR-control (Figure 1). The CR-removal reduced the 0–30 SMN levels relative to the control in the autumn after early-broccoli system, but not after the late-system in both years (Figure 1). No treatment differences were found in the 30–60 cm depth (data not shown), with SMN concentrations ranging from 17 to 51 kg N ha^{-1} in the CR-control.

Due to the lack of effect on SMN by the CC-oat compared to the CR-control (Figure 1, Table 3), it is suggested that oat cover crops may not reduce the potential for N losses after broccoli harvest. Conversely, the establishment of an oat cover crop after green pea production reduced autumn SMN concentrations in southwestern Ontario [41]. It is possible that the CC-oat had low N uptake compared to the plant available N in the soil after broccoli harvest, considering that 94 to 210 kg N ha^{-1} remained as available N at green pea harvest [41] while 265 to 415 kg N ha^{-1} remained at broccoli harvest.

Despite the removal of the N rich crop residue (CR-removal), the 0–30 cm SMN concentrations were only reduced in the autumn after the early-broccoli and not after the late-broccoli system (Figure 1). The difference in CR-removal effect between systems was likely a reflection of the cooler temperatures and slower N mineralization of the crop residue in the late-system. Regardless, high autumn SMN concentrations (89 to 227 kg N ha^{-1}) remained in the field in the CR-removal treatment (Figure 1). It is therefore suggested that the soil and/or below-ground crop residue provided substantial quantities of N during the post-harvest period. Thus, best management practices that minimize N losses after broccoli production would be beneficial.

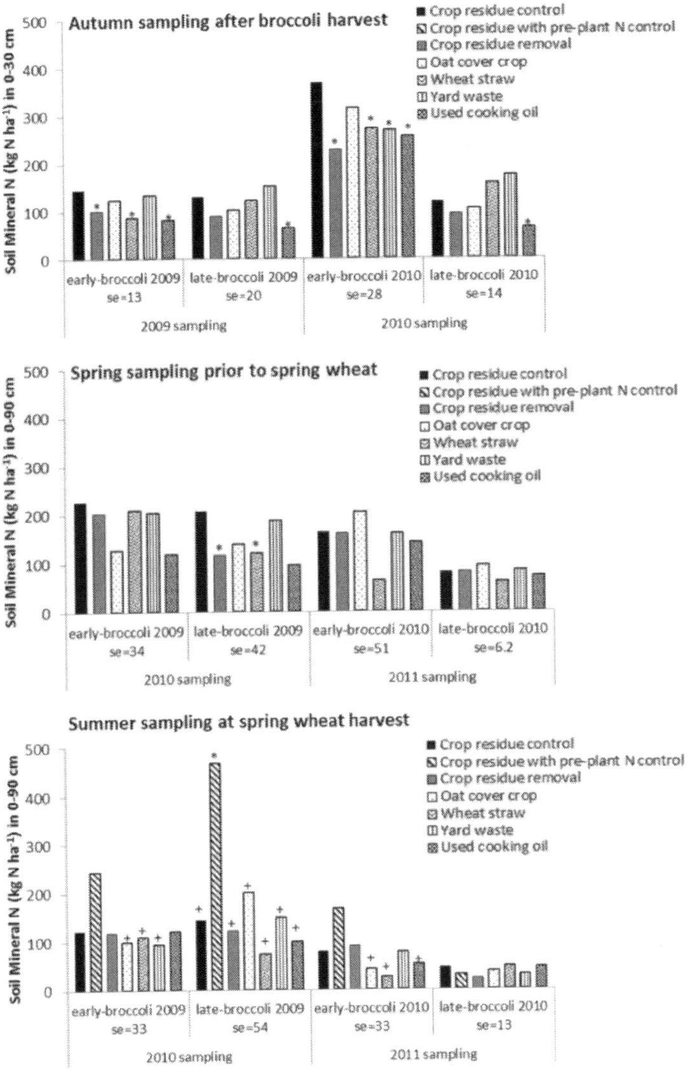

FIGURE 1: Soil mineral N concentrations in the autumn, spring, and summer after the 2009 and 2010 early- and late-broccoli harvest treatments. Symbols denote a significant difference ($P < 0.05$) compared to the crop residue control * or the crop residue with pre-plant N control +, based on a multiple means comparison with a Dunnett-Hsu adjustment. The se values represent the standard error of the mean.

Nitrogen immobilization is related to the biochemical composition of the decomposing substrate, and is positively associated with a high C:N ratio, lignin, and polyphenol content [42]. Incorporation of easily decomposable, high C:N ratio materials generally causes a rapid increase in microbial biomass and consequently SMN depletion as N is assimilated into microbial cells. Soil amendments such as yard waste, wheat straw, and used cooking oil/oily food waste have previously demonstrated N immobilization [11,19,23,25,26].

The OCA-yard reduced SMN compared to the CR-control only after the early-broccoli system in 2010 (Figure 1). The 2009 yard waste was composed of notably larger particles than that of 2010, thus it is possible that C decomposition took longer in 2009 and had less influence on microbial N immobilization. More recalcitrant substrates, such as the lignin-containing wood pieces of the OCA-yard, can have more limited decomposition at low temperatures than that of easily decomposed and more labile material [43]. It is likely that C decomposition of the OCA-yard was generally low, and consequently little microbial N immobilization occurred. Perhaps if the OCA-yard material was finely chopped, greater decomposition may occur. As opposed to the present study, previous laboratory research showed N immobilization by incorporating yard waste with broccoli crop residue [11], and field research has found that green waste compost mixed with cauliflower crop residue immobilized approximately 42 kg N ha^{-1} within the first month after incorporation [12]. Yet, in agreement with the present study, green waste composts have not resulted in N immobilization during autumn after the incorporation with cauliflower or leek residues in a two-year study [44]. Thus, the composition and substrate size of OCA-yard greatly influences N immobilization and its applicability as a better management practice for minimizing N losses in the autumn after broccoli harvest.

Although OCA-straw reduced SMN after the early-broccoli system, it must be noted that some wheat seed germinated and established a cover crop during both years. Thus, the reduction in autumn SMN may be a reflection of cover crop N uptake as well as microbial N assimilation. Conversely, it appeared that the OCA-straw treatment after the late-broccoli system did not sufficiently lower SMN, compared to the CR-control, to reduce the potential for soil N losses in the autumn. The difference in effect between systems was likely due to cooler temperatures and slower decomposition (or cover crop uptake) after late-broccoli. In comparison to previous laboratory research, it was demonstrated that wheat straw incorporated with broccoli crop residue could significantly lower SMN via N immobilization, (relative to incorporating crop residue alone) after 8 weeks of incubation [11]. Previously, cereal straw immobilized 35 kg N ha^{-1} within the first month after incorporation with cauliflower crop residues in the field [12]. It has been estimated that straw incorporation can result in the net N immobilization of 64 kg N ha^{-1} after two months [45], or 39 to 44 kg N ha^{-1} after one year [20], and reduce the amount of NO_3^--N leached by 27% after a year [45]. The current study has suggested that the OCA-straw treatment after the early-broccoli system can reduce SMN concentrations by 57 to 96 kg N ha^{-1}, which could otherwise be lost during autumn.

The OCA-oil treatment appeared to result in consistent N immobilization after both early- or late-broccoli systems (Figure 1). Compared to the typical grower practice (CR-control), the OCA-oil treatment had 53 to 112 kg ha^{-1} less SMN, thus 30% to 50% less SMN could be available for losses from the top 30 cm soil layer during the autumn after broccoli harvest. This finding was consistent with previous field research, which found that oily food waste application in autumn reduced soil NO_3^--N by immobilization and lowered the potential for N losses by 47 to 56 kg N ha^{-1} in the top 60 cm of soil [26]. Furthermore, a previous laboratory study suggested that OCA-oil immobilized more SMN than OCA-straw or OCA-yard, when incorporated with broccoli crop residue [11]. The rate of used cooking oil decomposition was synchronous with that of broccoli crop residue, whereas yard waste or wheat straw decomposed slower than broccoli crop residue or used cooking oil [11]. Because decomposition of more recalcitrant substances is more limited at low temperatures than that of easily decomposed material [43], the OCA-oil is likely the most promising material to reduce the potential for soil N losses due to its readily decomposable and labile matter. Thus, OCA-oil may be the most suitable amendment tested, because broccoli can be harvested anytime from early August to late October in southern Ontario.

3.3.3 SOIL MINERAL NITROGEN IN THE SUBSEQUENT SPRING AND SUMMER

The 0–90 cm profile was investigated to assess the quantity of SMN in the soil depth accessible for the crop, at spring wheat pre-plant and harvest. Prior to spring wheat planting, the SMN concentrations were generally not affected by the post-broccoli-treatments (Figure 1). Because the 2011 spring had a one and a half times higher precipitation (512 mm from February to May) compared to 2010 (343 mm from February to May), it is possible that SMN was subjected to different mechanisms of concentration reduction depending on the year. In the CR-control, NO_3^--N leaching may have occurred to a greater extent in 2011, thereby lowering spring SMN concentrations. Although NO_3^--N leaching could have also occurred in the OCA treatments, the amendments may have immobilized N in the spring, also lowering SMN concentrations. Additionally, the condition of high soil water content combined with the presence of a readily decomposable C source and SMN, denitrification could have been favored. Given the possibility of different mechanisms of SMN concentration reduction across treatments, further investigation is required. Future research should focus on 15N labeled crop residue to investigate the influence of the OCA treatments on the fate of the crop residue-derived N.

At spring wheat harvest, 0–90 cm SMN results indicated that the post-broccoli-harvest treatments did not lower SMN, compared to the CR-control (Figure 1). However, the application of N fertilizer at spring wheat pre-plant (CRN-control) left a larger quantity of SMN at harvest compared to most other treatments (Figure 1).

3.3.4 SPRING WHEAT PRODUCTION

Overall, the 2011 spring wheat grain contained 49 kg N ha^{-1} less N and had 1.6 Mg ha^{-1} less yield than 2010 (Figures 2 and 3). Early- and late-broccoli systems did not have different spring wheat yield or N contents. The yield ranged from 0.9 to 3.4 Mg ha^{-1} (Figure 3), similar to the average Ontario spring wheat yields of 3.5 Mg ha^{-1} in 2010 and 2011 [46]. The N harvest indices ranged from 55% to 90%, and did not vary by treatment (data not shown).

It must be noted that spring wheat plant N content, grain N content, plant biomass, or grain yield were never different between the CRN-control and the CR-control (Figures 2 and 3). It therefore appears that sufficient soil N was available for crop production subsequent to broccoli crop residue incorporation, regardless of pre-plant fertilizer application for spring wheat. Thus, growers may be able to reduce N fertilizer applications to spring wheat planting, following broccoli production because the mineralization of crop residue may provide sufficient SMN.

Relative to the CR-control, the treatments which indicated a detrimental effect on spring wheat production were the OCA-straw and CC-oat, based on a lower spring wheat plant N content, grain N content, plant biomass, and grain yield, compared to the CR-control or CRN-control (Figures 2 and 3). Also, the CR-removal treatment indicated a reduction in spring wheat plant biomass and grain yield compared to the CR-control or CRN-control in 2010 (Figures 2 and 3). Conversely, the OCA-yard and OCA-oil treatments did not have different plant N content, grain N content, plant biomass, and grain yield, compared to the CR-control or CRN-control (Figures 2 and 3). Therefore, it appears that the OCA-yard or OCA-oil treatments after broccoli harvest did not negatively impact the subsequent spring wheat production, but the OCA-straw, CC-oat, and CC-removal treatments resulted in spring wheat yield penalties.

It is possible that N supply was sufficient for spring wheat production in the CR-control, CRN-control, OCA-yard, and OCA-oil treatments. To optimize N use efficiency, early-season N availability has been recommended to achieve a desired spring wheat yield goal, rather than late-season N availability [27]. If early-season SMN levels were limiting for plant production, decreased vegetative dry matter accumulation and grain yield may occur [27]. Thus, the rate and timing of OCA and its decomposition is crucial for determining N dynamics. The grain yield results suggested that N supply was sufficient for the spring wheat growing season after the OCA-yard and OCA-oil, but perhaps not the OCA-straw (Figure 3). Likewise, the CC-oat and CR-removal treatments may not have had sufficient available N (Figure 3). Thus, N fertilizer applications may be required to maintain the spring wheat yield after OCA-straw, CC-oat, and CR-removal practices.

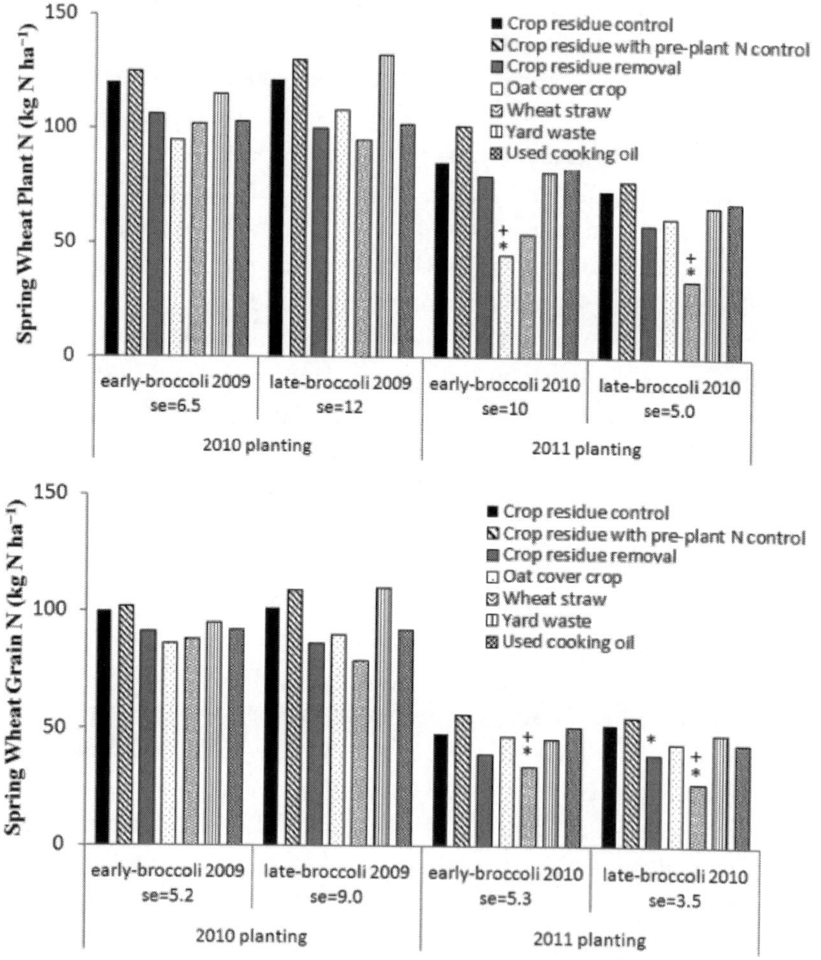

FIGURE 2: The N content (kg N ha^{-1}) of spring wheat plant biomass (grain + straw) and grain in 2010 and 2011, as affected by the previous years' treatments in the early- and late-broccoli systems. Symbols denote a significant difference ($P < 0.05$) compared to the crop residue control * or the crop residue with pre-plant N control +, based on a multiple means comparison with a Dunnett-Hsu adjustment. The se values represent the standard error of the mean.

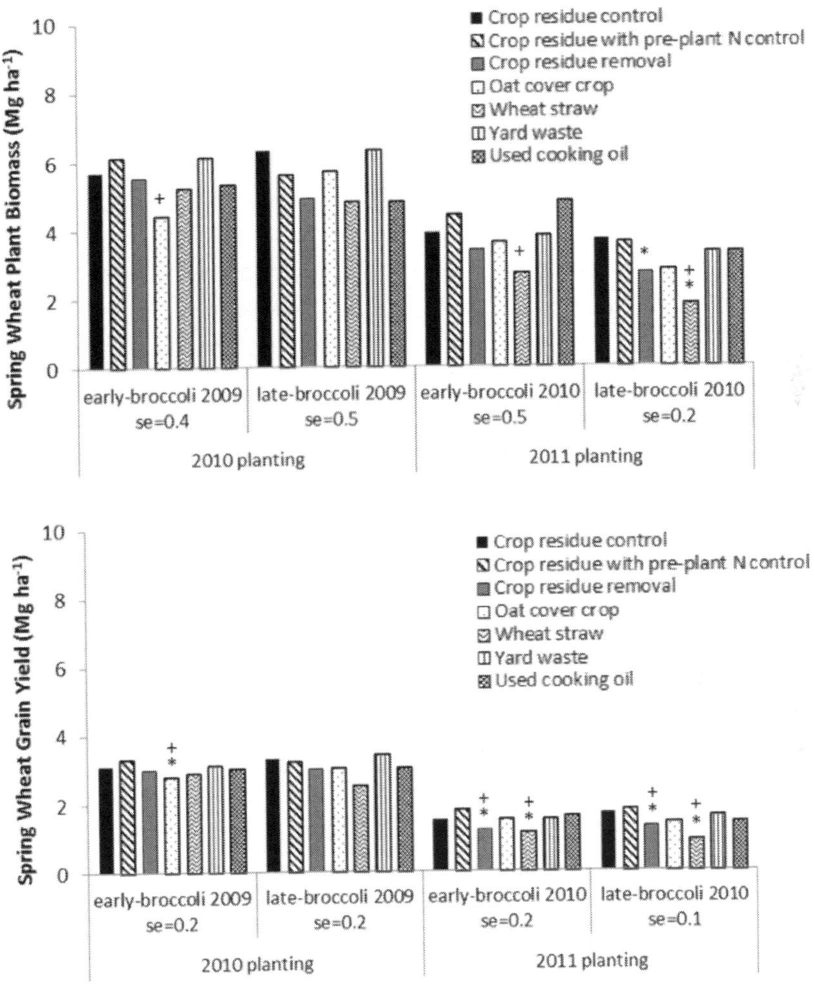

FIGURE 3: The spring wheat plant biomass (grain + straw) and grain yield (Mg ha^{-1}) in 2010 and 2011, as affected by the previous years' treatments in the early- and late-broccoli systems. Symbols denote a significant difference ($P < 0.05$) compared to the crop residue control * or the crop residue with pre-plant N control +, based on a multiple means comparison with a Dunnett-Hsu adjustment. The se values represent the standard error of the mean.

The lower spring wheat grain yield compared to the CR- or CRN-control in 2011 after the OCA-straw or CC-oat treatment (Figure 3) may have been an allelopathic result, because straw mulch often reduces subsequent wheat yields [47]. Considering that wheat seed in the OCA-straw treatment germinated and established a cover crop after broccoli harvest, the continuous cereal cropping from autumn to the subsequent summer may have accumulated phytotoxins or pathogens in the soil [47], which lowered the grain yield in 2011. Variation in environmental conditions between years likely contributed to the severity of phytotoxic or pathogenic factors. It is also possible that plant available N concentrations were not sufficient for optimal spring wheat yield in the OCA-straw and CC-oat treatments. In a previous study, which incorporated cereal straw with cauliflower crop residues, a pattern of autumn N immobilization was followed by N re-mineralization in the following spring [12]. However, a different study found no apparent re-mineralization even one year after straw incorporation in the field [20].

The lack of effect on spring wheat harvest parameters from the OCA-yard treatment (Figures 2 and 3) may be related to the limited effect on SMN after broccoli harvest in autumn (Figure 1). Similarly, limited autumn N immobilization and no consistent N re-mineralization were found following the autumn incorporation of green waste compost or sawdust with cauliflower or leek crop residues [12,44]. It was suggested that rye, leek, or lettuce production may not be negatively affected following autumn cauliflower crop residue incorporation with compost or straw amendment because plant N uptake or dry matter accumulation were generally similar between amendment and crop residue alone treatments [12], which was similar to the present OCA-yard results.

Considering that autumn N immobilization occurred in the OCA-oil treatment (Figure 1), yet spring wheat harvest parameters were generally not affected (Figures 2 and 3), it is possible that N losses may be reduced during a period of high risk for losses, without negatively affecting the subsequent crop. It is therefore suggested that the OCA-oil after broccoli harvest may be a better N management practice than the typical practice of the CR- or CRN-control. Although N was immobilized during the autumn after broccoli harvest, it appeared that OCA-oil did not conserve more N in the soil compared to the CR-control for subsequent spring wheat use

(Figure 1). A reason may be that little or no NO_3^--N leaching occurred in the CR-control treatment; the soil textures with 24% to 25% clay (loam or sandy clay loam) (Table 1) would support a slower rate of NO_3^--N leaching than sandier soils. Alternatively, the soil may have provided substantial quantities of plant available N regardless of the treatment, as suggested earlier. Thus, future experiments should investigate the fate of broccoli crop residue-derived N as influenced by the used cooking oil in N limited soil, sandy soil, or trace the crop residue-derived N into the subsequent crop via ^{15}N studies.

A pattern of N immobilization followed by re-mineralization with used cooking oil or "fat oil grease" amendments has been found in previous laboratory [11] and field [26] research. In the present study, N immobilized in the OCA-oil after broccoli harvest in autumn could have re-mineralized for the subsequent spring wheat plant uptake. Net N mineralization has also been observed in the spring following an autumn application of oily food waste [26]. Previous research has found corn yields to be similar between un-amended treatments and autumn amendments of "fat oil grease" [16]. Based on the maximum economic rate of N applied to the corn crop, it was found that N availability to corn was not affected by autumn application of "fat oil grease" [16]. The authors suggested that sufficient time was probably available for decomposition of the C material when it was applied in autumn [16]. Also, researchers have found little concern for detrimental accumulations of oily food waste and observed that the waste can promote water stable aggregation [25]. However, the spring application of "fat oil grease" prior to corn production resulted in net N immobilization during the spring [26], reduced corn yields up to 23%, and the additional requirement of 60 kg N ha^{-1} to offset corn yield declines [16]. Thus, "fat oil grease" or OCA-oil applied in the autumn did affect the following corn yield [26] or spring wheat yield.

3.3.5 ECONOMIC ANALYSIS

The potential environmental benefit of reduced N losses must be considered in combination with economic outcomes. Although the OCA-straw and CR-removal treatments showed some potential for reducing autumn

N losses after early-broccoli systems (Figure 1), it not only resulted in reduced spring wheat production parameters (Figures 2 and 3), but also reduced spring wheat profit margins (Table 3), relative to the CR- or CRN-control. Even without detrimental effects on spring wheat production parameters, the OCA-oil treatment reduced profit margins compared to the CR- or CRN-control (Figures 2 and 3, Table 3). However, of all the treatments which demonstrated a potential for reduced autumn N losses, (after early-broccoli by OCA-wheat, CR-removal treatments, or after both early- and late-broccoli by the OCA-oil), the OCA-oil treatment may be the least likely to lower spring wheat profit margins compared to CRN-control (Table 3). Furthermore, the economic results suggest that pre-plant application of N fertilizer for spring wheat was not necessary due to the similar profit margins between the CR-control and CRN-control (Table 3).

TABLE 3: The effect of post-broccoli-harvest treatments on 2010 and 2011 spring wheat profit margins ($ ha^{-1}) subsequent to the 2009 and 2010 early- and late-harvested broccoli.

Post-broccoli-harvest treatment	Broccoli production system			
	2009		2010	
	early-broccoli	late-broccoli	early-broccoli	late-broccoli
	Spring wheat profit margins ($ ha^{-1})			
	2010		2011	
Crop residue control	586	617	263	306
Crop residue with pre-plant N control	527	511	228	229
Crop residue removal	429 *,+	422 *	75 *,+	84 *,+
Oat cover crop	394 *,+	440 *	144 *,+	131 *,+
Wheat straw	434 *,+	352 *,+	70 *,+	24 *,+
Yard waste	515	579	205	228
Used cooking oil	508 *	506	235	197 *
Standard error of mean (se)	31.7	39.5	34.2	25.1

*Symbols denote a significant difference ($P < 0.05$) compared to the crop residue control * or to the crop residue with pre-plant N control +, based on a multiple means comparison with a Dunnett-Hsu adjustment.*

It has previously been indicated that trade-offs exist between environmental and economic benefits, because practices which reduced NO_3^--N losses caused lower economic returns [28]. Also, vegetable production practices, which reduced NO_3^--N leaching, did not always coincide with those that generated optimal economic outcomes [29]. Although a prior study on "fat oil grease" amendments did not specifically determine the economic impact [16], the impact can be estimated based on the observed effects on the following corn yield and N requirements. The reported mean control plot yields [16] and the average Ontario prices in 1996 and 1997 for corn and urea [48] suggested that the use of "fat oil grease" reduced the revenue by as much as \$290 ha^{-1} and increased the input costs by about \$52 ha^{-1}. The reduced profit margins of autumn applied OCA in corn was similar to the present study in a broccoli-spring wheat rotation. Thus, best management practices that have environmental benefit may also have an economic cost for growers.

3.4 CONCLUSIONS

Better N management practices are necessary after cole crop production due to the high SMN concentration and the risk for N losses in the post-harvest season. Although the OCA-straw demonstrated reduced autumn SMN concentrations after the early-broccoli system, it is not recommended if spring wheat is the following crop due to yield reductions. Also, the OCA-yard showed inconsistent potential for reduced autumn soil N losses. However, the practice of OCA-oil is recommended due to the reduced potential for autumn N losses via N immobilization after both early- and late-broccoli systems, without the subsequent spring wheat yield being detrimentally affected. This field study was consistent with previous incubation research, which found that OCA-oil demonstrated the most promise for the potential reduction in soil N losses after broccoli harvest [11]. Despite the environmental benefits of potentially reduced N losses by OCA, economic costs were associated. Thus, growers must evaluate the environmental vs. economic benefits of OCA-oil compared to the typical practice.

REFERENCES

1. Bakker, C.J.; Swanton, C.J.; McKeown, A.W. Broccoli growth in response to increasing rates of pre-plant nitrogen. II. Dry matter and nitrogen accumulation. Can. J. Plant Sci. 2009, 89, 539–548.
2. Thompson, T.L.; Doerge, T.A.; Godin, R.E. Subsurface drip irrigation and fertigation of broccoli: I. Yield, quality, and nitrogen uptake. Soil Sci. Soc. Am. J. 2002, 66, 186–192.
3. De Neve, S.; Hofman, G. N mineralization and nitrate leaching from vegetable crop residues under field conditions: A model evaluation. Soil Biol. Biochem. 1998, 30, 2067–2075.
4. Everaarts, A.; de Willigen, P. The effect of the rate and method of nitrogen application on nitrogen uptake and utilization by broccoli (Brassica oleracea var. italica). Neth. J. Agric. Sci. 1999, 47, 201–214.
5. Fallow, D.J.; Brown, D.M.; Parkin, G.W.; Lauzon, J.D.; Wagner-Riddle, C. Identification of Critical Regions for Water Quality Monitoring with Respect to Seasonal and Annual Water Surplus; Land Resource Science Technical Report 2003-1 for Ministry of Agriculture and Food, Ontario Agricultural College Department of Land Resource Science, University of Guelph: Guelph, ON, Canada, 2003.
6. Zebarth, B.J.; Bowen, P.A.; Toivonen, P.M.A. Influence of nitrogen fertilization on broccoli yield, nitrogen accumulation and apparent fertilizer-nitrogen recovery. Can. J. Plant Sci. 1995, 75, 717–725.
7. Everaarts, A.; de Willigen, P. The effect of nitrogen and the method of application on yield and quality of broccoli. Neth. J. Agric. Sci. 1999, 47, 123–133.
8. Yoldas, F.; Ceylan, S.; Yagmur, B.; Mordogan, N. Effects of nitrogen fertilizer on yield quality and nutrient content in broccoli. J. Plant Nutr. 2008, 31, 1333–1343.
9. Bakker, C.J.; Swanton, C.J.; McKeown, A.W. Broccoli growth in response to increasing rates of pre-plant nitrogen. I. Yield and quality. Can. J. Plant Sci. 2009, 89, 527–537.
10. De Neve, S.; Pannier, J.; Hofman, G. Temperature effects on C- and N-mineralization from vegetable crop residues. Plant Soil 1996, 181, 25–30.
11. Congreves, K.A.; Voroney, R.P.; O'Halloran, I.P.; Van Eerd, L.L. Broccoli residue-derived nitrogen immobilization following amendments of organic carbon: An incubation study. Can. J. Soil Sci. 2013, 93, 23–31.
12. Chaves, B.; de Neve, S.; Boeckx, P.; van Cleemput, O.; Hofman, G. Manipulating nitrogen release from nitrogen-rich crop residues using organic wastes under field conditions. Soil Sci. Soc. Am. J. 2007, 71, 1240–1250.
13. Oo, A. Availability for Bio-processing in Ontario, 2012; Technical Report for Ontario Federation of Agriculture: Guelph, Canada, 2012. Available online: http://www.ofa.on.ca/uploads/userfiles/files/biomass_crop_residues_availability_for_bioprocessing_final_oct_2_2012.pdf (accessed on 1 October 2012).
14. Field Crops: Area and Production Estimates. Ministry of Agriculture, Food and Rural Affairs Web site, 2010. Available online: http://www.omafra.gov.on.ca/english/stats/crops/ctywwheat10.htm (accessed on 1 October 2012).

15. Van der Werf, P. State of Organic Waste Diversion in Ontario; Technical Report for Ontario Waste Management Association: Brampton, Canada, 2009. Available online: http://www.2cg.ca/pdffiles/State%20of%20Composting%20in%20Ontario%200509-1.pdf (accessed on 1 October 2012).
16. Rashid, M.T.; Voroney, R.P. Land application of oily food waste and corn production on amended soils. Agron. J. 2004, 96, 997–1004.
17. Organic Resource Management Incorporation (ORMI) Web site. Available online: http://www.ormi.com/ormi/ (accessed on 4 February 2013).
18. Horticultural Crops: Area, Production, Value and Sales by Crop. Ministry of Agriculture, Food and Rural Affairs Web site, 2011. Available online: http://www.omafra.gov.on.ca/english/stats/hort/veg_all10-11English.pdf (accessed on 1 October 2012).
19. Jawson, M.; Elliott, L. Carbon and nitrogen transformations during wheat straw and root decomposition. Soil Biol. Biochem. 1986, 18, 15–22.
20. Mary, B.; Recous, S.; Darwis, D.; Robin, D. Interactions between decomposition of plant residues and nitrogen cycling in soil. Plant Soil 1996, 181, 71–82.
21. Bending, G.D; Turner, M.K. Interaction of biochemical quality and particle size of crop residues and its effect on the microbial biomass and nitrogen dynamics following incorporation into soil. Biol. Fertil. Soils 1999, 29, 319–327.
22. Hartz, T.; Giannini, C. Duration of composting of yard wastes affects both physical and chemical characteristics of compost and plant growth. HortScience 1998, 33, 1192–1196.
23. Amlinger, F.; Götz, B.; Dreher, P.; Geszti, J.; Weissteiner, C. Nitrogen in biowaste and yard waste compost: dynamics of mobilisation and availability—A review. Eur. J. Soil Biol. 2003, 39, 107–116.
24. Claassen, V.; Carey, J. Regeneration of nitrogen fertility in disturbed soils using composts. Compost Sci. Util. 2004, 12, 145–152.
25. Plante, A.F.; Voroney, R.P. Decomposition of land applied oily food waste and associated changes in soil aggregate stability. J. Environ. Qual. 1998, 27, 395–402.
26. Rashid, M.T.; Voroney, R.P. Recycling soil nitrate nitrogen by amending agricultural lands with oily food waste. J. Environ. Qual. 2003, 32, 1881–1886.
27. Cassman, K.G.; Bryant, D.C.; Fulton, A.E.; Jackson, L.F. Nitrogen supply effects on partitioning of dry matter and nitrogen to grain of irrigated wheat. Crop Sci. 1992, 32, 1251–1258.
28. Saleh, A.; Osei, E.; Jaynes, D.B.; Du, B.; Arnold, J.G. Economic and environmental impacts of LSNT and cover crops for nitrate-nitrogen reduction in Walnut Creek watershed, Iowa, using FEM and enhanced SWAT models. Trans. ASABE 2007, 50, 1251–1259.
29. Thompson, T.L.; Doerge, T.A.; Godin, R.E. Nitrogen and water interactions in subsurface drip-irrigated cauliflower: II. Agronomic, economic, and environmental outcomes. Soil Sci. Soc. Am. J. 2000, 64, 412–418.
30. Pier, J.W.; Doerge, T.A. Concurrent evaluation of agronomic, economic, and environmental aspects of trickle irrigated watermelon production. J. Environ. Qual. 1995, 24, 79–86.

31. Thompson, T.L.; Doerge, T.A. Nitrogen and water interactions in subsurface trickle-irrigated leaf lettuce: II. Agronomic, economic, and environmental outcomes. Soil Sci. Soc. Am. J. 1996, 60, 168–173.
32. Carter, M.R.; Gregorich, E.G. Soil Sampling and Methods of Analysis, 2nd ed.; CRC: Boca Ratan, FL, USA, 2008.
33. OMAFRA. Agronomy Guide for Field Crops; Queen's Printer for Ontario: Toronto, Canada, 2009.
34. Rutherford, P.M.; McGill, W.B.; Arocena, J.M.; Figueiredo, C.T. Total Nitrogen. In Soil Sampling and Methods of Analysis; Carter, M.R., Gregorich, E.G., Eds.; CRC: Boca Ratan, FL, USA, 2008; pp. 239–250.
35. Maynard, D.G.; Kalra, Y.P.; Crumbaugh, J.A. Nitrate and Exchangeable Ammonium Nitrogen. In Soil Sampling and Methods of Analysis; Carter, M.R., Gregorich, E.G., Eds.; CRC: Boca Ratan, FL, USA, 2008; pp. 71–80.
36. Field Crop Prices, by Crop Year, 1981–2010. Ontario Ministry of Agriculture, Food and Rural Affairs Web site, 2010. Available online: http://www.omafra.gov.on.ca/english/stats/crops/price_swheat.htm (accessed on 20 June 2012).
37. Survey of Custom Framework Rates Charged in 2009. Ontario Ministry of Agriculture, Food and Rural Affairs Web site, 2009. Available online: http://www.omafra.gov.on.ca/english/busdev/facts/10-049a1.htm (accessed on 20 June 2012).
38. Ontario Farm Input Monitoring Project, Survey No. 1, 2009, 2010, 2011; Technical Report for Economics and Business Group, University of Guelph: Ridgetown, Canada, 2011.
39. Gutezeit, B. Yield and nitrogen balance of broccoli at different soil moisture levels. Irrig. Sci. 2004, 23, 21–27.
40. Bowen, P.; Zebarth, B.; Toivonen, P. Dynamics of nitrogen and dry-matter partitioning and accumulation in broccoli (Brassica oleracea var. italica) in relation to extractable soil inorganic nitrogen. Can. J. Plant Sci. 1999, 79, 277–286.
41. O'Reilly, K.A.; Lauzon, J.D.; Vyn, R.J.; Van Eerd, L.L. Nitrogen cycling, profit margins and sweet corn yield under fall cover crop systems. Can. J. Soil Sci. 2012, 92, 353–365.
42. Cabrera, M.L.; Kissel, D.E.; Vigil, M.F. Nitrogen mineralization from organic residues: Research opportunities. J. Environ. Qual. 2005, 34, 75–79.
43. Nicolardot, B.; Fauvet, G.; Cheneby, D. Carbon and nitrogen cycling through soil microbial biomass at various temperatures. Soil Biol. Biochem. 1994, 26, 253–261.
44. Chaves, B.; de Neve, S.; Piulats, L.; Boeckx, P.; van Cleemput, O.; Hofman, G. Manipulating the N release from N-rich crop residues by using organic wastes on soils with different textures. Soil Use Manag. 2007, 23, 212–219.
45. Garnier, P.; Néel, C.; Aita, C.; Recous, S.; Lafolie, F.; Mary, B. Modelling carbon and nitrogen dynamics in a bare soil with and without straw incorporation. Eur. J. Soil Sci. 2003, 54, 555–568.
46. Estimated Area, Yield, Production and Farm Value of Specified Field Crops, Ontario, 2001–2012. Ontario Ministry of Agriculture, Food and Rural Affairs Web site, 2012. Available online: http://www.omafra.gov.on.ca/english/stats/crops/estimate_metric.htm (accessed on 2 August 2012).
47. Wu, H.; Pratley, J.; Lemerle, D.; Haig, T. Allelopathy in wheat (Triticum aestivum). Ann. Appl. Biol. 2001, 139, 1–9.

48. Ontario Farm Input Monitoring Project, Survey No. 1–4, 1996–1997; Technical Report for Economics and Business Group, University of Guelph: Ridgetown, Canada, 1997.

CHAPTER 4

SUSTAINABILITY OF US ORGANIC BEEF AND DAIRY PRODUCTION SYSTEMS: SOIL, PLANT AND CATTLE INTERACTIONS

AIMEE N. HAFLA, JENNIFER W. MACADAM , AND KATHY J. SODER

4.1 INTRODUCTION: ORGANIC VS. NON-ORGANIC CATTLE FEEDING PRACTICES

In the United States, organic agricultural systems differ from non-organic systems in many important ways, most notably in restrictions on the use of chemical fertilizers and pesticides in all organic farming systems, and in restrictions on the use of antibiotics, feed additives and growth hormones in organic animal systems. The availability and cost of certified organic feeds is a major challenge for US organic dairy and beef producers. In 2011, NASS estimated that 83,989,000 acres of corn were harvested in the US, with only 134,877 of those certified organic. So little organic corn and soybean production occurs that limited supply results in prices of certified organic feedstuffs that are two to three times greater compared to non-organically grown feeds [1]. The cost of organic supplements for energy and protein results in more on-farm production of these feeds in organic

This chapter was originally published under the Creative Commons Attribution License. Hafla AN, MacAdam JW and Soder KJ. Sustainability of US Organic Beef and Dairy Production Systems: Soil, Plant and Cattle Interactions. Sustainability **5** *(2013); 3009-3034. doi:10.3390/su5073009.*

systems compared to typical non-organic systems. However, some farms are too small or have land that is undesirable for the production of annual crops and therefore may have to rely on purchasing these feeds. Moreover, availability of local organic feeds can be limited, adding transportation expenses to the cost of the feed. In 2011, the state of Iowa was the greatest producer of certified organic corn, while California had the greatest inventory of certified organic dairy and beef cattle [2]. Locally available by-product feeds resulting from grain processing, production of human foods and beverages and other manufacturing enterprises can provide livestock producers with lower cost feed supplements that provide protein and energy to support the growth and lactation of ruminant livestock. Some common by-product feeds utilized in non-organic dairy and feedlot rations include oil meals, bran, middlings, brewers grains, distillers grains and beet pulp [3]. However, the availability of by-products produced to meet the National Organic Program (NOP) certified organic standards are limited and therefore organic producers are unlikely to have access to these feedstuffs to reduce feed costs. In the 2011 Certified Organic Production Survey conducted by the USDA-NASS, the quantity and value of commodity crops, fruits, vegetables and livestock produced were summarized; however, no organically produced by-products appeared in the report.

4.1.1 THE PASTURE RULE

In 2010 the NOP implemented the Pasture Rule, requiring that certified organic ruminant animals consume pasture during the grazing season in their geographic region [4]. More specifically, the rule requires that during the grazing season animals must obtain an average minimum of 30% of their dry matter intake (DMI) by grazing certified organic pasture or rangeland and must graze at least 120 days each year (not necessarily continuous). Ruminant livestock that are to be grain finished are exempt from the requirement of 30% DMI from pasture during the finishing period, but must have access to pasture; however, the finishing period is not to exceed one-fifth of the animal's total life or 120 days [4]. The intent of the pasture rule was to establish specific enforceable standards regarding access to pasture by ruminant livestock and to assure consumers of organic products that livestock are raised in pasture-based systems, which are considered more natural and humane than confinement systems. Previous to implementation of the pasture rule, organic producers of dairy and beef cattle were required to provide animals with access to pasture that was managed for feed value; however, no frequency or duration of grazing time or proportion of grazed forage in the diet was specified. The implementation of the pasture rule has leveled the playing field between small and large certified organic dairy and meat producers by increasing the use of grazing in all organic ruminant enterprises, diminishing the economies of scale for the very largest producers.

The organic food production sector continues to grow [5] but less than 2% of US dairy production is currently organic. While statistics on non-organic use of pasture for dairy production are difficult to document, smaller non-organic dairy producers in the upper Midwest and Northeast are the most likely to rely on pasture as feed [6]. In general, non-organic dairy production and beef finishing are predominantly confinement systems in the US while organic beef and dairy production rely more on pasture. As of 2011, there was approximately 1.6 million acres of certified organic pasture and rangeland on 3,499 farms in the United States [2]. This paper focuses on the implications of the use of pasture within the restrictions that apply to organic dairy and beef agricultural systems in the US.

4.1.2 DAIRY FEEDING PRACTICES

While smaller scale non-organic dairies may use pasture as a significant part of their feed, the majority of non-organic dairy cows are fed a combination of supplemental protein and energy sources (purchased or homegrown) and homegrown conserved forages such as silage, haylage, and dry hay. Organic dairy farms, on the other hand, rely more on pasture during the grazing season [7,8] with some supplemental feeds (energy and/or conserved forages) as needed. In a review by Benbrook [9], it was found that the overall reliance on pasture for all non-organic dairy farms in the US was 3% of DMI, and that 75% of cows on non-organic farms did not rely on pasture for any part of the lactating cow diet. However, it is important to note that in much of the US, climatic conditions require the harvest and storage of feeds on organic dairy farms to supply animals during drought and winter months.

4.1.3 BEEF FEEDING PRACTICES

While production practices vary across regions, most beef cattle intended for commercial beef production spend a majority of their lives on grasslands and are subsequently finished with grain in commercial feedlots. The greatest divergence between organic and non-organic beef feeding practices is therefore during the finishing phase. Approximately 80% of non-organically raised beef cattle in the US are finished using grain-based (usually corn) rations [10]. Conversely, in a survey of certified organic beef producers in the US, 83% reported that they raised cattle exclusively or predominantly on grass and hay until slaughter, while the remaining 17% reported using a grain finishing system [11]. All feeds consumed by certified organic ruminants must comply with NOP standards; therefore, all grains and conserved forages must be 100% certified organic. Vitamins, trace minerals and feed additives that are not produced organically are allowed in the diets of organic livestock in trace amounts provided they are allowed under the National List of Allowed and Prohibited Substances [12]. Finally, growth promotants such as ionophores and hormones [13] and antimicrobials [14] that are commonly used in beef feedlot finishing are not allowed in organic beef finishing.

4.2 ORGANIC PASTURE MANAGEMENT

The USDA definition of pasture is land that is used for livestock grazing that is managed to provide feed value and maintain or improve soil, water, and vegetative resources [12]. More specifically, the goals for pastures managed for organic production of livestock are to build soil structure and fertility, manage pests ecologically, conserve and promote plant and animal biodiversity, and maximize forage quality and production. Because organic and pasture-based livestock producers must rely on pastures managed without chemical fertilizers or herbicides for milk production or live weight gain, they are likely to place greater emphasis on pasture management than non-organic producers.

Managed grazing is not exclusive to organic livestock producers; however, due to the pasture rule organic dairy producers are far more likely to use intensive grazing management practices than are non-organic dairy operations. Rotational grazing and strip grazing are ruminant management strategies common to both organic dairy and beef farms, in regions and seasons where environmental and forage conditions allow. Rotational stocking is a method that uses recurring periods of grazing and rest among three or more pastures [15]. Rotational stocking can ensure intake of high quality forage for livestock throughout the grazing season, distribute forage growth optimally over time, and lengthen the grazing season [16]. Strip stocking is a variation of rotational stocking that may employ annual crops such as cereal grains or summer annuals (e.g., triticale, Austrian winter pea, sorghum-sudangrass) that are cultivated for grazing. Animals progress across a field, allotted a specific amount of feed for a short time, controlling feed intake and optimizing feeding efficiency [15]. Rotational and strip grazing are two of the most common grazing management practices utilized in grazing ruminant production systems; however, there are many other specific management practices in use across various regions.

4.2.1 SOIL FERTILITY IN ORGANIC AGRICULTURAL SYSTEMS

In a meta-review of soil and environmental quality studies conducted in New Zealand, Switzerland, and the Great Plains of the US, Condron et al. [17] determined that soil organic matter and soil biological activity were consistently higher in organic systems compared to non-organic systems. Soil structure, cation exchange capacity, respiration and the soil content of mineralizable nitrogen relative to organic carbon were higher in organically farmed New Zealand soils, as was the number of earthworms [18]. However, soil pH and the content of phosphorus and sulfur were lower. This is consistent with other studies cited by Condron et al. [17], such as Liebig and Doran [19] and Nguyen et al. [20], who demonstrated that the imbalance between outputs and inputs, which were minimal for phosphorus and sulfur in the organic systems evaluated, lead to deficiencies over time. Løes and Øgaard [21] studied the change in plant available phosphorus over time in organic systems, and determined that maintenance of soil phosphorus is essential for the long-term productivity of organic systems. Cornforth and Sinclair [22] demonstrated that in grazing systems, the higher the stocking rate, the greater the need for phosphorus inputs. Gosling and Shepherd [23] also demonstrated the depletion of another important nutrient, potassium, under long-term organic farming. Where long-term manuring has increased plant-available potassium to undesirably high levels, nutrient export from farms in the form of beef and dairy products will have a long-term benefit. In the northeastern US, where heavy manure application has occurred for decades, nutrients such as potassium (and even phosphorus) have built up to very high levels. It is not unusual for pastures in the Northeast to have potassium levels in excess of 3%, high enough to cause milk fevers [24]. The time required for plants to access phosphorus mineralized from organic sources such as manure [25] and from allowed mineral sources [17] compared with synthetic phosphate fertilizers varies greatly, which means that long-term planning is required for the proper maintenance of soil fertility in organic production systems.

Since the use of synthetic fertilizers is prohibited in organic systems, producers must rely on animal manures, compost and cover crops to increase soil nitrogen content. Nitrogen is needed in large quantities and is often a limiting nutrient in organic pastures and crop fields. In non-organic systems this issue is addressed by the use of nitrate or ammonia fertilization. However, nitrogen in the nitrate (NO_3^-) form is not held by the soil and can leach below the root zone into groundwater, even when it is applied as split applications, two or three times within the growing season, to minimize application in excess of plant need, and nitrate is a serious contaminant of drinking water. Microbes can convert nitrogen into forms that volatilize from the soil; nitrogen volatilized as nitrous oxide (N_2O) has 300 times the greenhouse gas effect of carbon dioxide (CO_2). A fundamental understanding among organic practitioners is that nutrients should be recycled within each farm and the transfer of nutrients, leading to depletion or excess across farm boundaries, should be minimized. Manure is one of the best sources of nitrogen available to organic producers; manure also contains approximately 90% water by weight, so it is challenging to transport or distribute. Therefore, manure is commonly disposed of by applying it to the soil. However, drylot dairies and feedlots import nutrients and must compost manure and/or export it to surrounding cropland. While cropping-based (e.g., hay, silage, and cereal grains) dairy feeding operations often import nutrients to the farm and distribute the resulting manure nutrients on nearby cropland two or more times per year, pasture-based livestock feeding returns nutrients to the pasture during grazing [26].

Pastured ruminants are an important conduit for the return of the nutrients that are not exported from the farm as beef or dairy products to pasture soils. Grazing-based ruminant production allows most of the nutrients from a pasture to be immediately recycled onto the same pasture, reducing the loss of volatile nutrients (e.g., ammonia, methane) from manure storage and compost piles and lagoons, and reducing the time and fuel required to mechanically spread manure solids onto cropland. However, the distribution of manure on pastures is uneven at best. The nitrogen in urine patches is sufficiently concentrated that nitrate may leach from pastures into groundwater, forage that is soiled by manure will be rejected for a period of time, and unincorporated manure solids are vulnerable to runoff. Of the macronutrients, calcium, phosphorus and magnesium are largely excreted in the feces, while potassium is largely secreted in urine. Sulfur secretion is variable, but more nitrogen is excreted in urine than in feces, although this is influenced by diet. The form of nitrogen found in urine is readily converted to a form that can be volatilized by an enzyme present in feces, while nitrogen is excreted in feces as organic matter, which is mineralized much more slowly by soil microbial activity [27]. The management of cattle on pastures can influence the distribution of manure nutrients; rotational stocking or other types of management-intensive grazing (MiG) such as strip grazing result in a more even distribution of nutrients than continuous grazing [28]. Under rotational stocking, reasonably well-distributed manure nutrients can contribute to a flush of soil microbial activity in pastures, particularly if pastures are irrigated following grazing, and nutrients can be extracted and utilized by growing forage plants during the rest period between grazing events. The grazing pressure exerted by higher stocking densities or co-grazing with other species can reduce rejection of forages adjacent to urine or dung patches.

Cederberg and Mattsson [29] modeled Swedish non-organic and organic dairy nutrient inputs and outputs, and determined that while there were nutrient surpluses generated by both systems, they were one-tenth as high for phosphorus and potassium, and one-third as high for nitrogen on organic dairies. The need for nitrogen fertilization can be reduced by including nitrogen-fixing legumes, but bloat-causing legumes such as the *Trifolium* (true) clovers are limited to no more than 50% of mixtures. Benbrook [9] estimated that in 2008, due to the organic dairy industry alone, 40 million pounds of synthetic nitrogen was not applied to land designated for certified organic use. Integrating livestock and cropping enterprises on a single farm not only spreads economic risk, but can improve the efficiency of nutrient cycling. Mixed crop-livestock systems such as dairies that grow their own conserved feed for winter, have been shown to minimize inputs due to synergies between components, specifically fertilizer, putting them at lower risk for nitrogen pollution than other systems [30]. It is well accepted that legume-based systems can reduce nitrogen losses associated with cropping [31] and grazing [32]; therefore, organic producers utilizing mixtures of grass and legume forage species are able to take advantage of naturally occurring nitrogen fixation.

4.2.2 SOIL ORGANIC MATTER

A number of soil factors can influence forage plant growth and nutrient acquisition, which becomes more critical as the proportion of forages in a ruminant‘s diet increases. If soil bulk density is excessive, plant root growth will be negatively affected, and if soil becomes dry, the dissolution and movement of nutrient ions in the soil solution will be restricted. The pH of the soil can also affect the availability of nutrients, some to toxic levels when soils become very acidic, and the proliferation of soil microbes, including those involved with legumes in nitrogen fixation. While high plant productivity can be achieved on unhealthy soils through the additional of high levels of fertilizer, long-term soil health is a goal of organic agricultural systems because these —conventional‖ options are not available in organic systems [31].

A healthy soil is a soil with an optimal pH, aeration, and nutrient- and water-holding capacity considering the parent material, plant community and climate under which it was formed. A healthy agricultural soil is one that is managed sustainably to improve deficits in any of these traits in support of plant growth without degradation, such as erosion or salinization. One of the most important indicators of a healthy soil is the level of organic matter. Like other carbon-based fuels, organic matter is consumed more rapidly by soil microbes at high temperatures and high levels of oxygen, so while good drainage and aeration are important for root growth, the cultivation required to support annual cropping hastens the breakdown of organic matter. Chemical fertilization stimulates the microbial mineralization of soil organic carbon [33], but the greater agricultural productivity achieved with synthetic nitrogen fertilization can result in higher crop residues and increase soil organic matter. However, high levels of nitrogen and phosphorus fertilization can undermine soil stability and lead to increased erosion [34].

Organic matter in the soil is a repository not only of carbon, but of the nitrogen and phosphorus compounds contained in the flora and fauna of which it is comprised. Soil organic matter as humus glues soil particles together, creating structure that is resistant to collapse and therefore is essential to drainage and aeration, soil organic matter has more surface area and a far greater capacity to absorb water than the mineral components of the soil [35], so it not only adsorbs nutrient cations (the soil cation exchange capacity) but can hold nearly its own weight in soil water. Since the quantity of water used by a given plant species is directly proportional to its leaf growth [36], and therefore to its vegetative productivity, the ability of a soil to absorb and release water is a critical trait.

4.2.3 THE IMPACT OF GRAZING MANAGEMENT ON FORAGE PLANTS

To the extent that grazed or conserved forages comprise a greater amount of the feed of organic ruminants than non-organically fed ruminant livestock, the health and performance of these animals may be significantly influenced by the mineral composition of forages. The uptake of nutrients by forages is a function of rooting depth, root surface area, and forage plant species. In general, the root length and density of perennial forages is greater than that of annual crops such as soybean, corn or the cereal grains such as wheat or barley [27]. The effective rooting depth of perennial forage legumes may be greater (e.g., alfalfa) or less (e.g., white clover) than that of grasses, and legumes often store carbohydrates and proteins for regrowth underground, in the tap root and crown (the overwintering stem base).

Pasture management can have a profound effect on root function as well as on the botanical composition of pastures: In grasses, the fibrous root systems are comprised of many relatively thin roots, and grasses have a relatively shallow rooting depth. Overgrazing grasses will deplete both the aboveground shoot material as well as the carbohydrate storage of the grass plant, which is located in the stem or leaf bases near the soil surface (stubble), while well-managed grazing of grazing-tolerant forage species can enhance the soil by encouraging the turnover of fibrous roots, increasing soil organic matter. Excessive removal of photosynthetic tissue slows the rate of regrowth of the shoot, which is largely dependent on current photosynthesis to support growth. Overgrazing also results in self-pruning of the root system, because the root has no capacity for photosynthesis and is therefore entirely dependent on current or stored photosynthesis to support the maintenance of existing roots, much less growth. Under continuous grazing, where ruminants have access to a large area of pasture or rangeland for months at a time, the most desirable forage species will be grazed first, and will be re-grazed as soon as they begin regrowth. This undermines the ability of the shoot to compete for sunlight and of the root to compete for water and soil nutrients. Over time, the botanical composition of grazingland will shift away from the most desirable plant species to the least desirable species, which were left ungrazed and able to successfully compete for resources, reproduce vegetatively by tillering, stolons or rhizomes, and complete their life cycle by maturing and distributing seed.

4.3 THE EFFECT OF GRAZING ON RUMINANTS

Ruminants that acquire the majority of their feed from forages, as hay or by grazing, may be more affected by variations in forage nutrient quality and availability than those on diets that contain higher proportions of supplemental feeds. Further, ruminants may selectively absorb or passively excrete mineral nutrients and nitrogen, altering the effect of dietary mineral content. Whitehead [27] notes five homeostatic control mechanisms that help to maintain ruminant nutrient uptake within a healthy range:

1. Variation in the extent of absorption from the digestive tract.
2. Variation in the extent of excretion via the feces.
3. Variation in the extent of excretion via the urine.
4. Deposition of an excess in the tissues in harmless and/or reserve forms.
5. Variation in the extent of secretion into milk.

For dairy cows that excrete significant quantities of some mineral nutrients in milk, dietary concentrations may be critical. While there is a metabolically active reserve of calcium and magnesium in bone, the reserve amounts are small relative to intake or excretion. With such a small buffer, a dietary deficiency can rapidly have serious negative or even fatal effects for a cow early in lactation. Therefore the maintenance of a healthy soil with good soil structure to encourage extensive root systems, good water-holding capacity to buffer drought or suboptimal irrigation, and a capacity to provide and retain good soil fertility is fundamental to the sustainability of grazing-based ruminant livestock systems, particularly in pastures regularly used by dairy cows that excrete high levels of mineral nutrients and nitrogen on a daily basis.

4.3.1 FORAGE PLANT NUTRITION IN PASTURES

Plant roots, like soils, have characteristic cation exchange capacities (CEC), influenced by the concentration of carboxylic acids in root cell walls that provide negative charges, especially if the pH of the soil solution is a unit or two below neutrality (pH 5–6) [27]. In general, legume roots have a higher CEC than grasses, and therefore can compete more successfully for divalent cations, such as Ca^{2+} or Mg^{2+}, in the soil solution; among legumes, white clover accumulates more calcium than other legume species [27]. The balance of cations in the soil can affect their uptake, because much of the uptake of soil nutrients by roots is passive—

ions move with water in the soil solution, and move into roots as part of the transpiration stream. Uptake of nutrients that are passively acquired from the soil solution may be higher in mid-summer, when evapotranspiration is high. In contrast, nitrate, phosphate and sulfate are all negatively charged, and therefore are repelled rather than attracted to the negative charges on soil particles and roots.

An excess of potassium (K^+) in the soil solution, which is more likely in heavily manured soils, can result in —luxury uptake‖ by forages in excess of plant requirements, and lead to an imbalance of high potassium and therefore a relative deficiency of calcium and magnesium [37]. High K^+ concentrations in forages can result in metabolic issues in grazing ruminants due more to antagonisms with other elements such as calcium (Ca) and magnesium (Mg), than to simple deficiencies. Grass tetany (hypomagnesemia) is characterized by low blood Mg and usually occurs in beef cows and ewes grazing lush grass pastures during periods of cool and cloudy weather [38–40]. Grass tetany tends to occur when the dietary intake of total Mg is not particularly low, but instead when factors exist that increase the animal‘s Mg requirement (early lactation) or reduce the availability of Mg to the animal. Potassium inhibits the animals‘ ability to absorb Mg resulting in a relative deficiency, and therefore the ratio of milliequivalents of K to (Ca+Mg) in forage has been used to predict the tetany hazard of forages [39]. Furthermore, milk fever (hypocalcemia) is characterized by low levels of blood calcium and most commonly occurs in high producing dairy cows 12 to 24 h after calving when the sudden demand for Ca required by the onset of lactation tests the Ca homeostatic capabilities of the animal [41]. Excessive K concentrations in the prepartum diet of high-producing dairy cows decreases Ca resorption from the bone resulting in an imbalance in Ca homeostasis and a possible deficiency regardless of dietary concentrations of Ca [41]. Soder and Stout [37] noted that the high K concentrations observed in orchardgrass pastures fertilized with a dairy manure slurry could predispose lactating dairy cows to milk fever. Hardeng and Edge [42] examined the incidence of disease between 31 organic and 93 non-organic Norwegian dairy herds and unexpectedly found no differences between the production systems for cases of milk fever. The authors noted that for each kg increase in peak milk production, the risk of milk fever increased by 5%; however, the mean maximum milk production for organic herds was 4.6 kg lower compared to non-organic herds. Therefore, they suggested that the lower milk production found in the organic herds resulted in reduced Ca depletion from milking compared to non-organic herds [42].

Nitrate toxicity (methaemoglobinaemia) in ruminants grazing forages grown under stressful conditions or on soils that have received high applications of either inorganic (synthetic) or organic (manure) fertilizers and can be an issue in both organic and non-organic grazing systems. Plants take up nitrogen from the soil largely in the form of nitrate (NO_3^-), which under normal growing conditions is rapidly converted to nitrite (NO_2^-), then to ammonia (NH_3) and finally to plant proteins. However, when plant growth is slowed by stressful environmental conditions (frost, drought, shade), nitrogen in the form of NO_3^- accumulates in the plant more rapidly than it can be converted to protein. Nitrate itself is not toxic; however, microbial action on NO_3^- in the rumen results in the conversion of hemoglobin to methemoglobin, which greatly reduces the oxygen carrying capacity of the blood [39]. Nitrate levels remain stable after forage is harvested and cured for hay, but potential for accumulation is variable across forage species, with annual cereals (sorghum-sudan, sudangrass, oats, etc.) and some weeds (pigweed, lambsquarter, smartweed, etc.) documented as nitrate accumulators [38].

Forage micronutrient content is affected by soil content and plant uptake. The availability of the micronutrients iron, zinc, cobalt, manganese and copper is greater in acidic soils, while the availability of molybdenum and selenium is higher in alkaline soils. Iron is primarily used by ruminants to enable hemoglobin to carry oxygen in the blood for respiration; a deficiency can lead to anemia. Manganese, molybdenum, selenium and copper are needed for enzyme activity, and cobalt is a component of vitamin B-12. Zinc is required for enzyme activity, but is also needed to stabilize RNA and DNA, and for membrane function [27]. Deficiencies of manganese are uncommon, but the primary symptom is lameness. Deficiency symptoms of zinc and copper are not specific, but include poor growth in young stock and increased susceptibility to disease. Copper deficiencies are relatively common, and reduced copper absorption can be caused by high dietary molybdenum, sulfur, or zinc. Cobalt deficiencies result in loss of appetite, and are common either where soils are naturally low in cobalt, or where high levels of iron or manganese reduce the availability of cobalt to plants. Excesses of iron or zinc can interact to reduce the absorbance of each other [43]. Selenium is required for the enzyme glutathione peroxidase, and a deficiency of selenium can result in excessive lipid peroxidation, called white muscle disease, which can be fatal in young sheep and cattle. Excessive selenium can also be deadly: most locoweed plants are selenium accumulators, and their consumption can produce acute selenium toxicity, leading to rapid death; chronic consumption of selenium accumulator plants can lead to blindness or infertility [27].

4.3.2 VITAMINS AND MINERALS IN ORGANIC CATTLE SYSTEMS

Vitamins and minerals are provided to dairy cows in a non-organic system by including them in the total mixed ration (TMR). By mixing vitamins and minerals in a formulated ration, the animals cannot choose to consume more or less of a specific ingredient and selective feeding is limited. The practices of organic dairy producers can include a similar vitamin and mineral supplement using allowed ingredients, but may be more varied in regard to delivering vitamins and minerals. Organic dairy feeding systems range from large scale organic dairies (more common in the western US), where a mixed ration (partial TMR) is used along with required pasture, to small-scale exclusively pasture-based dairies (dominant in the upper Midwest and the northeastern US). When TMRs are not used, vitamin and minerals may be provided for free-choice consumption by animals. Additionally, providing individual minerals (cafeteria style) to allow for a truly free-choice selection instead of mixing or including them in a TMR has been suggested as a way to allow individual animal selection to meet nutritional requirements when forage quality is variable [44]. Beef cattle on pasture and rangeland may be provided free-choice access to mixed vitamin and minerals, a practice that is similar in non-organic and organic beef cow-calf systems. Mixed organic swards containing plants of white clover, chicory and plantain have been found to have greater mineral concentrations compared to perennial ryegrass or timothy [45], as do weeds such as dandelion, dock and chickweed [27]. Soder et al. [46] reported that increasing the biodiversity of a basic orchardgrass-white clover pasture increased mineral concentration of the pasture, particularly when chicory was included in the mixture. Restrictions on herbicide use and the greater biodiversity often found in organic pastures may result in greater concentrations of weeds in organic systems. Therefore, the mixed pastures often found in organic systems may help provide livestock with needed minerals through plant species biodiversity. Similar to the non-organic feed industry, within the organic sector there is a wide range of products marketed to improve animal performance and health. Kelp meal is commonly used as a natural source of chelated minerals throughout the northeastern US on organic dairy farms; however, limited scientific information is available on the impacts of including kelp meal in dairy or beef rations. Allen et al. [47] reported that feeding kelp extract to grazing beef cattle reduced body temperature and increased cell-mediated immune function during hot weather. However, others have reported no benefits to kelp feeding of dairy cattle for long periods of time [48] or short term [49].

4.3.3 TREATMENT OF PARASITES IN ORGANIC CATTLE SYSTEMS

Parasitic infections of ruminants as a result of grazing are considered one of the greatest challenges to animal welfare in organic production systems [50]. In non-organic systems, anthelmintics are routinely used to control or prevent heavy infestations of internal parasites. However, this practice is prohibited under the NOP regulations except in extreme cases of heavy infestation [4]. While fluke and lungworm infections can be addressed with some success by vaccinations which are allowed under National Organic Program rules in the US, nematodes continue to pose problems in organic systems. Temperate forages containing condensed tannin (e.g., birdsfoot trefoil, sainfoin, and sulla) have been found to have efficacy against nematodes [50,51]. Alternative dewormers such as garlic and diatomaceous earth have shown little efficacy in scientific studies despite anecdotal claims [52,53]. Commonly recommended practices to control internal parasite issues include: (1) graze youngstock on the cleanest pastures (pastures not grazed by animals of the same species within the past year, or pastures that have hay harvested); (2) don't overgraze, which forces animals to graze close to the ground where the highest concentration of parasite larvae reside; (3) utilize rotational grazing; (4) implement multi-species grazing; (5) where possible, break up parasite cycles by making hay at least once annually; and (6) healthy animals improves resistance to parasites, so provide adequate nutrition, clean water, and a balanced trace mineral mix [54,55]. In addition, genetic selection of animals for resistance or resilience to internal parasites may play a more important role in future animal production systems as decreased use of chemicals and increased anthelmintic resistance become more prominent.

4.4 GRAZING IMPACTS ON THE NUTRITIONAL QUALITY OF MILK AND MEAT

Evidence of improved nutrient content in fruits and vegetables produced under organic conditions is conflicting. Some studies have reported increased nutritional quality in organically raised produce when compared to non-organic produce [56,57], while others have found no differences between production systems [58,59]. However, impacts on the nutritional quality of ruminant products from cattle in pasture-based systems compared with non-organic confinement systems have proven to be more quantifiable.

4.4.1 QUALITY DIFFERENCES OF DAIRY PRODUCTS

Milk and other dairy products are a substantial source of saturated fatty acids (SFA) in the human diet and these fats significantly contribute to the risk of cardiovascular disease [60,61]; however, the value of dairy products as an important source of protein, vitamins, and minerals cannot be ignored. Therefore, improving the fat composition of dairy products by decreasing SFA content and increasing fatty acids that have positive effects on human health would greatly benefit consumers. It is well accepted that fatty acid composition and concentration in milk are dependent on animal physiology (breed, age, stage of lactation) [62,63] and diet [64,65]. Alpha-linolenic acid (ALA) is the primary dietary precursor of omega-3 fatty acids (FA) and conjugated linoleic acid (CLA) in the milk of cattle, and can account for as much as 50–70% of the total fatty acids found in fresh forage [66,67]. Therefore, it is no surprise that milk from cows grazing fresh grass contains greater concentrations of omega-3 FA and CLA compared to milk from cows consuming conserved forage, corn silage and high-grain rations [64,68–70]. A recent meta-analysis examining 29 previously published studies from 12 countries concluded that organic dairy products contained significantly higher protein, ALA, total omega-3 FAs, CLA and a more desirable omega-3 to omega-6 ratio compared to dairy products derived from non-organic systems [71]. Increasing omega-3 FAs in the human diet is considered beneficial to health and CLAs have been shown to be a powerful anticarcinogen in animal models [72].

In 2011, milk sold as certified organic in the United States reached nearly 1.2 billion kg and was valued at 7.6 million US dollars [2], representing an ever growing segment of the dairy industry. The implementation of the pasture rule insures that 30% of dry matter feed intake by certified organic ruminants will be obtained by grazing certified organic pasture or rangeland during at least 120 days of the year. Improvements in the fatty acid profile of organic dairy products, including greater omega-3 FA and CLA concentrations and more desirable omega-3 to omega-6 FA ratios will result from the organic grazing requirement. This is evident in England, where organic dairy production standards require a reliance on forage, and greater concentrations of omega-3 FA, CLA, and fat-soluble antioxidants have been found in organic milk compared to milk from dry-lot systems [73,74]. Seasonal variation in the fatty acid profile of milk produced under grazing conditions is a challenge associated with producing and marketing products that may be considered enriched with beneficial fatty acids. Concentrations of nutritionally beneficial fatty acids have been found to be greater in raw and retail milk during the grazing season compared to periods when the animals were consuming a greater proportion of grain and conserved forages [74,75]. Supplementing the diets of milking cows with oilseeds such as flax may be a method of elevating desirable milk fats when fresh forage is limited by seasonality or drought [69].

4.4.2 QUALITY DIFFERENCES OF BEEF PRODUCTS

The impacts of forage feeding on production and quality of the end-products of beef cattle are varied. It is generally accepted that steers finished in a forage-based system have leaner, lighter carcasses (or must spend more time on feed to reach equivalent carcass weights) and may produce beef with a greater intensity of off-flavor, compared to those from grain-based systems [76–79]. Some direct comparisons of grain- and grass-finished beef have reported undesirable eating characteristics (specifically tenderness and juiciness) of forage finished beef, but these studies were confounded by age at slaughter or plane of nutrition effects [80–82]. Duckett et al. [77] compared forage (alfalfa, pearl millet or mixed pasture) and grain-based (corn-silage) finishing diets on carcass and meat quality of Angus-cross steers when finished to an equal time point. As expected, the authors found forage-finished steers to have 46% less total fat on the 9th to 11th rib section. No differences in meat quality and eating quality parameters, including juiciness and overall tenderness, were reported between the finishing systems; however, a trained sensory panel found a tendency for the forage-finished beef to have a greater intensity of off-flavor [77].

Increasing the vitamin E content of beef, specifically through grass feeding, has been suggested as a potential method to stabilize some of the negative sensory and storage effects associated with altered fatty acid content of forage finished beef [83]. Further research has focused on the impacts of forage finishing on altering fatty acid composition of beef products, specifically reducing SFA content, improving the omega-6 to omega-3 FA ratio, enhancing CLA content and increasing B-vitamins and antioxidants. Beef from cattle produced in forage-based systems have a lower overall fat content and greater omega-3 FA and CLA content compared to beef from cattle finished using grain-based diets [77,83–85]. Moreover, vitamin E and β-carotene concentrations have been found to be greater in beef finished on pasture compared to traditional finishing programs [84]. However, it is important to note that while the concentrations of beneficial fatty acids are greater in forage-finished beef, the total fat content is lower, and therefore may impact the total amount (as mg/d) of beneficial fatty acids in whole cuts of beef. Additionally, while forage feeding increases some beneficial fatty acids that have been prominent topics in current human health trends (e.g., omega-3 FA and CLA), the impact on other fats and their potential for human health benefits have received less attention. Concentrate-finishing enhances oleic acid concentration in beef [77,86], and diets high in mono-unsaturated fatty acids (MUFA), specifically oleic acid, can lower the concentrations of undesirable low-density-lipoprotein cholesterol (LDL) humans [87]; low-density-lipoprotein cholesterol represents a significant risk factor for cardiovascular disease and are considered the bad cholesterol [88]. More research is needed on the direct human health effects of consuming beef products considered enriched with omega-3 FA and CLAs (forage fed) or with oleic acid (concentrate-finished).

Consumers are willing to pay a premium for animal products from organic systems because they should not contain residues of the synthetic chemicals (pesticides, herbicides, fertilizers, fungicides and veterinary drugs) that are used in non-organic ruminant production systems [89]. However, the nutritional benefits from organic milk and meat products in the form of healthier fat composition and increased antioxidant and vitamin properties are due to dietary factors that are used in but not exclusive to organic systems, specifically greater pasture consumption of organic dairy and beef cattle compared to those in non-organic systems. However, grazing-based dairy and beef systems (including certified organic systems) do not have exclusive claims to these greater levels of potentially beneficial components in meat and milk products. It is possible to increase beneficial fatty acids and other nutritional components of meat and milk in confinement feeding through the use of supplemental feeds such as oilseeds [90,91] or fish oil [92].

4.5 GREENHOUSE GAS IMPLICATIONS OF ORGANIC AND NON-ORGANIC PRODUCTION

4.5.1 GREENHOUSE GAS EMISSIONS IN DAIRY SYSTEMS

Greenhouse gas (GHG) emissions from organic dairy systems have been found to be lower on a per-ha area than emissions from non-organic dairy systems across a range of environments in western Europe, based on whole-farm assessments using the FarmGHG model [93]. Life cycle assessments (LCA) that compared organic and non-organic dairy systems in Scandinavia [29] and in the Netherlands [94] also identified advantages in organic dairy production. Lynch et al. [95] noted the scarcity of US and Canadian data for GHG emissions associated with dairy systems. A US comparison of organic, non-organic, and non-organic plus recombinant bovine somatotropin (rbST) dairy production, which included the GHG value of measured inputs, outputs and land area, concluded that organic systems had the highest environmental impact [96]. However, this study focused on the efficiency of production per cow and was neither a whole-farm study nor a LCA; therefore, the implications of related factors, such as culling rate, were not considered. Even the LCA and whole-farm modeling studies did not include the economic or social values of these systems or their ecosystem services.

According to a Food and Agriculture Organization LCA of the dairy sector, US milk production represents about 16% of the world total, while the relative contribution of the GHG emissions, per kg of fat- and protein-corrected milk (FPCM), associated with milk production, processing and transportation in the US, is about 8% of the world total. This ratio of GHG emissions to milk production is the lowest of any global region, including western Europe [97] (Figure 4.2). Within this highly efficient US dairy sector, organic accounts for only about 4% of production [98], but organic dairy sales have continued to increase by double digits, even during the Great Recession that began in 2008 [99].

4.5.2 GREENHOUSE GAS EMISSIONS IN BEEF SYSTEMS

While the per-head carbon footprint of grain-finished cattle was higher as estimated by Peters et al. [100] in an Australian study, it was offset by high feedlot rates of gain. However, the contribution of legumes in pastures was not determined in that study, and in a review of greenhouse gas emissions from beef and dairy cattle production systems, no reference was made to forage legumes [101]. There is an opportunity to reassess these beef LCA incorporating MiG and high-quality forages that can greatly increase the rate of gain of cattle finished on pasture. Evidence for this comes from studies of plants used for biofuel production. Kim and Dale [102] reported that alfalfa, a perennial, N-fixing legume, resulted in the lowest g carbon-equivalent per-kg biomass emissions among corn (246–286), soybeans (159–163) and switchgrass (124–147) compared with just 89 for alfalfa. Finishing beef cattle on high quality pastures can reduce GHG emissions and the global warming potential (GWP) associated with grass finished beef. This can occur through increased sequestration of CO_2 into soil organic carbon and/or through decreased emissions of N_2O and methane with their potent GWP [103,104]. Rangelands and pastures are well-known for their ability to increase soil carbon stocks [105], and this trait should differ little between non-organic and organic systems.

Beef cattle production conducted on rangeland generally requires a significant amount of land, and therefore tends to be extensive and uncultivated with intact natural biodiversity [106]. There are approximately 2 million acres of certified organic rangeland in the US [107], but little information is available on the implications of management practices specific to certified organic rangeland, much of which is found in the western US. Because chemical fertilization and weed control are usually not economically feasible on non-organic rangeland, organic rangeland would not be expected to differ greatly from non-organic. In a review by Sayre et al. [106], it was noted that ranching conducted on rangelands exemplifies many of the defining characteristics of diversified farming systems, as it relies on the functional diversity of natural processes of both plants and animals and creates services generated and regenerated locally without nonrenewable inputs.

4.6 ECONOMICS OF GRAZING-BASED CATTLE SYSTEMS

Feed is the greatest cost associated with milk and meat production [108]. Therefore, when grazed forage can replace the use of more expensive grains or conserved forages it is possible to decrease the input costs of ruminant production systems.

4.6.1 *ECONOMICS OF DAIRY SYSTEMS*

Many studies have conducted direct comparisons of pasture-based (but not necessarily organic) and confinement dairy systems and have reported that producers utilizing grazing have lower milk production per cow but also have lower operating expenses and greater net incomes per cow compared with confinement systems [108–112]. In a 4-year study, White et al. [108] examined economic measures in pasture systems using Holstein and Jersey cows and found that cows in a pasture-based system produced 11% less milk than cows in confinement. However, feed costs were on average 31% lower for the grazing cows compared to cows in confinement, and income over feed was found to be the same between the two systems. Within organic systems, McBride and Greene [113] noted that while feed costs per cow were 25% less for organic dairies that used pasture for the majority of their feed compared with organic dairies that used the least pasture, milk production on pasture-intensive organic dairies was 30% less than for dairies with minimal pasture use. However, this study was conducted before the implementation of the pasture rule in 2010 which will have significantly increased the pasture feeding of organic cows previously fed like non-organic dairy cows but with organically sourced feeds.

In general, economic factors including labor for animal care, manure handling, forage management and cow culling rates were more desirable for pasture-based systems. Veterinary costs were lower for organic than non-organic dairy farms in the Netherlands [114] and the health of organic dairy cows was found to be better than that of non-organic dairy cows [115]. Organic dairy cows spend more time grazing, which results fewer hoof and leg problems in part due to spending less time on concrete, and organic dairy producers are more inclined to administer approved remedies or take a wait-and-see approach before contacting a veterinarian [42]. Hanson et al. [111] examined farm income tax returns from 62 dairy farms in the mid-Atlantic region to evaluate financial performance of dairy producers utilizing either MiG or confinement production systems. The authors found that purchased feed expenses were 31% less in the MiG systems compared to confinement systems (267 vs. 387 $/acre, respectively). Additionally, MiG operations produced less milk but had substantially lower costs of production and were more profitable than confinement operations per hundredweight of milk produced and on a per-cow basis. Moreover, there was no impact on total farm profitability between the production systems.

Grazing-based dairy systems are the dominant dairy production model in New Zealand, Australia and the European Union [116,117]; however, not all grazing-based dairies are organic. On a world scale, Dillon et al. [118] found a strong inverse relationship between costs of production and proportion of grazed forage in the cow‘s diet, and reported that Australia and New Zealand had the greatest proportion of the diet provided by grass (85%–90%) and the lowest cost of production of all the countries evaluated. Additionally, the authors noted that increasing the proportion of grazed grass in a system where pasture was already the predominant feed will reduce the cost of production. For example, extending the grazing season by 27 days in Irish systems reduced the cost of milk by 1 cent/L [118]. In Europe, pasture is generally accepted as the cheapest feed source for dairy cattle and forms the base for profitable low-input systems in this region [119].

Reliable studies directly comparing the economics of organic and non-organic milk production in the US are limited, and those studies conducted previous to the 2010 pasture rule may not apply to the economics of organic systems today. Butler [120] compared production costs of organic and non-organic dairy farms in California and found the total cost of milk production per cwt (45 kg) was 16% higher in the organic system due largely to greater feed, labor, herd replacement and transition costs. However, the authors acknowledged that feed costs, which make up half of total production costs, were not as high as expected in the organic systems due to the utilization of low-cost pasture. A study by Rotz et al. [121] demonstrated that organic dairies typical of the northeastern US were more profitable than comparably sized non-organic dairies, but that profitability depended on a premium price for organic milk.

4.6.2 ECONOMICS OF BEEF SYSTEMS

The impact of grazing on profitably of the beef industry as a whole is complicated given the segmentation of a typical animal's life into a cow-calf suckling/grazing period, a growing/developing period for replacement stock or a growing/fattening period for slaughter stock. More than

70% of the lifetime body weight of a beef animal destined for slaughter is spent on forages and up to 100% of the life of beef breeding stock may be spent on pastures or conserved forages [122]; however, the degree of grain supplementation varies by region. The cow-calf sector represents a part of the industry where animals spend a significant amount of time on pasture and rangeland and generally require a minimal amount of purchased feed prior to finishing. Therefore, this phase may already be similar to practices of organic systems [123] and adjustments to adhere to organic standards would be more focused on eliminating the use of implantable growth promotants and antibiotics rather than feeding and housing.

When evaluating costs of production during the growing and finishing phase, Fernandez and Woodward [124] found that the cost of gain was greater for calves in an organic system compared to calves fed in a non-organic system ($1.89 vs. $ 1.36/kg gain, respectively). In non-organic beef finishing systems the major determinant for cost of gain is the price of feed, and in this study 79% of the cost of gain was due to feed costs for the organically raised steers, where 73% of the cost of gain was due to feed for the non-organically raised steers, with these costs heavily dependent on the price of organic and non-organic corn. This study was conducted before the implementation of the pasture rule and did not address profit, which would include premiums for the sale of organically raised cattle. However based on cost of gain, the organic steers cost 39% more to finish than the non-organically raised steers, when grain-finishing was used [124].

Increasing corn and soybean prices due to expanding demand of these commodities for agrofuel production have impacted both non-organic and organic ruminant production systems that depend on high-grain feeds to maximize production. The price of organic feed corn increased from an average of $5.22/bu in 2010 to $10.68/bu in 2011 [98] causing many livestock producers to explore alternative feed sources in an effort to become less dependent on cereal grains. Moreover, increased input costs from energy-dense feeds like cereal grains caused some producers to evaluate the economics of pasturing calves in stocker systems or forage finishing beef in an effort to reduce cost of gain. Lewis et al. [125] evaluated the economics of an intensive beef production system, where cattle were weaned and immediately finished on a high-grain diet compared with an extensive system where calves were weaned, wintered on crop residues, grazed on summer pastures and then finished on a high-grain diet. The break-even price for cattle in the extensive system was significantly lower due to the additional body weight added to the animals from grazing forage before entering the feedlot. Cattle in the extensive system produced more total beef at a lower cost per unit of product. The profitability of the extensive system was due largely to the low cost of forage feeding.

Currently, few reliable direct comparisons have been made regarding profitability between organic and non-organic ruminant production systems in the United States; however, increased utilization of grazed forages is generally associated with reduced feed costs and lower costs of production per animal in both dairy and beef systems. Therefore, the use of grazing to meet the requirements of the NOP pasture rule may subsequently reduce input costs and contribute to economic profitability in organic ruminant production systems more than premiums from the organic label itself.

4.7 SOCIAL IMPLICATIONS OF ORGANIC VS. NON-ORGANIC CATTLE PRODUCTION SYSTEMS

Organic agriculture is founded on a set of principles that guide the technical approach to plant and animal production, but which also accept responsibility for the environmental and social consequences of food production. The four principles of organic agriculture, as articulated by the International Federation of Organic Agriculture Movements, are as follows:

The Principle of Health—Organic agriculture should sustain and enhance the health of soil, plant, animal, human and planet as one and indivisible.

The Principle of Ecology—Organic agriculture should be based on living ecological systems and cycles, work with them, emulate them and help sustain them.

The Principle of Fairness—Organic agriculture should build on relationships that ensure fairness with regard to the common environment and life opportunities.

The Principle of Care—Organic agriculture should be managed in a precautionary and responsible manner to protect the health and well being of current and future generations and the environment.

The social implications of beef and dairy production practices are more difficult to quantify than production or economics, but the welfare of farm workers is predicted to be the —next chapter of the food movement‖ by Michael Pollan [126]. The effects of organic livestock production on rural communities, land stewardship and human health have already been considered:

Rural Communities

Organic agriculture strives to be sustainable and therefore to protect the whole of the environment [127], including workers, owners, and the rural communities to which they belong. The greater use of pasture in organic dairy and beef production systems due to the pasture rule has the potential to influence farm size, which may be more important than farm type in predicting community participation and involvement. The premium that consumers are willing to pay for organically produced food allows small, labor-intensive businesses to prosper. Additionally, the use of locally available inputs is encouraged in organic agriculture, which in turn increases demand for other local businesses. Sustaining smaller organic dairies and beef producers by enforcing the pasture rule and supporting a premium price for organic beef should result in more farm residents, which should benefit communities [128]. However, the total number of organic producers within dairy or ranching communities is small, making the impact of this factor difficult to quantify.

Land Stewardship

Protecting and enhancing soil fertility by adding manure, compost or by growing and plowing under cover crops is important in organic agriculture, not only to improve the current land productivity, but also to ensure production for future generations. Where perennial pastures replace cropping systems, there will be a decrease in soil erosion, cultivation and harvesting, and an increase in organic matter and (where nitrogen-fixing legumes are used) an increase in the nitrogen content of the soil, supporting productivity and the cycling of mineral nutrients through forage use and the return of nutrients to the soil as urine and dung.

Human Health

Organic dairy and beef products are not contaminated with pesticide, hormone or antibiotic residues, and have been shown to have improved nutritional content, including increased beneficial nutrients such as omega-3 fatty acids and conjugated linoleic acids [129]. A strongly positive public perception of the perceived benefits results in increased consumption of organic foods by consumers.

There is increasing interest in organic and sustainable cattle production systems by both producers and consumers, which can be seen in the continued willingness of consumers to pay more for organic products and in the continued expansion of the organic beef and dairy agricultural sectors to meet this need. However, organic production still constitutes too small a segment of dairy and beef production for the potential impacts on rural communities, farm families, and farm workers to be quantified.

4.8 CONCLUSIONS

One of the most significant differences between organic and non-organic US dairy and beef production systems is the greater use of pasture in US organic systems. There are also non-organic grazing-based dairies as well as grass-finished beef enterprises that have more in common with US organic dairy and beef systems than with large-scale commercial non-organic operations. Among the implications of feeding cattle on pasture for a significant portion of the year are differences in manure management and soil organic matter, and greater vulnerability of cattle to certain nutrient deficiencies and parasites. It is more challenging to provide some mineral nutrients for organic crop production, so organic beef and dairy systems require longer-term nutrient planning and are more likely than non-organic systems to develop deficiencies in phosphorus, potassium and sulfur. However, because nitrogen is supplied from livestock manure or green manure in organic systems, soil organic matter tends to be higher in organic than non-organic systems. The economic well-being of US organic beef and dairy systems depends on the willingness of consumers to pay a premium for organically produced livestock products, and the increasing cost of nitrogen fertilizer and cereal grains as livestock feed creates a further incentive for beef and dairy producers to increase the use of pasture. The differences in the fatty acid composition of organic milk and meat are well-documented, and consumer preference for purchasing livestock products that have been produced without the use of antibiotics or growth hormones continues to increase; therefore, certified organic agriculture continues to be the fastest growing sector of US agriculture.

REFERENCES

1. Marston, S.P.; Clark, G.W.; Anderson, G.W.; Kersbergen, R.J.; Lunak, M.; Marcinkowski, D.P.; Murphy, M.R.; Schwab, C.G.; Erickson, P.S. Maximizing profit on New England organic dairy farms: An economic comparison of 4 total mixed rations for organic Holsteins and Jerseys. J. Dairy Sci. 2011, 94, 3184–3201.
2. United States Department of Agricultue, National Agricultural Statistics Service (USDA-NASS). 2011 Certified Organic Production Survey; United States Department of Agriculture-National Agricultural Statistics Service: Washington, DC, USA, 2012.
3. Mirzaei-Aghsaghali, A.; Maheri-Sis, N. Nutritive value of some agro-industrial by-products for ruminants—A review. World J. Zool. 2008, 3, 40–46.
4. Rinehart, L.; Baier, A. Pasture for Organic Ruminant Livestock: Understanding and Implementing the National Organic Program (NOP) Pasture Rule; United States Department of Agriculture, National Center for Appropriate Technology, National Sustainable Agriculture Information Service (ATTRA): Washington, DC, USA, 2011.
5. Sato, K.; Bartlett, P.C.; Erskine, R.J.; Kaneene, J.B. A comparison of production and management between Wisconsin organic and conventional dairy herds. Livest. Prod. Sci. 2005, 93, 105–115.
6. Gillespie, J.; Nehring, R.; Hallahan, C.; Sandretto, C. Pasture-based dairy systems: Who are the producers and are their operations more profitable than conventional dairies? J. Agr. Resour. Econ. 2009, 34, 412–427.
7. Hoshide, A.K.; Halloran, J.M.; Kersbergen, R.J.; Griffin, T.S.; DeFauw, S.L.; LaGasse, B.J.; Jain, S. Effects of stored feed cropping systems and farm size on the profitability of Maine organic dairy farm simulations. J. Dairy Sci. 2011, 94, 5710–5723.
8. Zwald, A.G.; Ruegg, P.L.; Kaneene, J.B.; Warnick, L.D.; Wells, S.J.; Fossler, C.; Halbert, L.W. Management practices and reported antimicrobial usage on conventional and organic dairy farms. J. Dairy Sci. 2004, 87, 191–201.
9. Benbrook, C. Shades of Green: Quantifying the Benefits of Organic Dairy Production; The Organic Center: Washington, DC, USA, 2009.
10. Matthews, K.H.; Johnson, R.J. Alternative Beef Production Systems: Issues and Implications. United States Department of Agriculture, Economic Research Service.

LDPM-218-01: 2013. Available online: http://www.ers.usda.gov/media/1071057/ldpm-218-01.pdf (accessed on 2 July 2013).

11. Cornucopia Institute. Position Paper on Organic Beef Finishing and Proposal for Three Tiered Labeling System for Organic Meat from Ruminants; Cornucopia Institute: Cornucopia, WI, USA, 2010.
12. Coffey, L.; Baier, A.H. Guide for Organic Livestock Producers; United States Department of Agriculture, National Center for Appropriate Technology, National Sustainable Agriculture Information Service (ATTRA): Butte, MT, UDA, 2012.
13. Kolok, A.S.; Sellin, M.K. The Environmental Impact of Growth-Promoting Compounds Employed by the United States Beef Cattle Industry: History, Current Knowledge, and Future Directions. In Reviews of Environmental Contamination and Toxicology; Whitacre, D.M., Ed.; Springer: New York, NY, USA, 2008; pp. 1–30.
14. Rogers, S.; Haines, J. Detecting and Mitigating the Environmental Impact of Fecal Pathogens Originating from Confined Animal Feeding Operations: Review; EPA/600/R-06/021: 2005; United States Envrionmental Protection Agency, National Risk Management Research Laboratory, Office of Research and Development: Cincinnati, OH, USA, 2005.
15. Allen, V.G.; Batello, C.; Berretta, E.J.; Hodgson, J.; Kothmann, M.; Li, X.; McIvor, J.; Milne, J.; Morris, C.; Peeters, A.; et al. An international terminology for grazing lands and grazing animals. Grass Forage Sci. 2011, 66, 2–28.
16. Walton, P.D.; Martinez, R.; Bailey, A.W. A comparison of continuous and rotational grazing. J. Range Manage. 1981, 34, 19–21.
17. Condron, L.M.; Cameron, K.C.; Di, H.J.; Clough, T.J.; Forbes, E.A.; McLaren, R.G.; Silva, R.G. A comparison of soil and environmental quality under organic and conventional farming systems in New Zealand. New Zeal. J. Agr. Res. 2000, 43, 443–466.
18. Reganold, J.P.; Palmer, A.S.; Lockhart, J.C.; Macgregor, A.N. Soil quality and financial performance of biodynamic and conventional farms in New Zealand. Science 1993, 260, 344–349.
19. Liebig, M.A.; Doran, J.W. Impact of organic production practices on soil quality indicators. J. Environ. Qual. 1999, 28, 1601–1609.
20. Nguyen, M.L.; Haynes, R.J.; Goh, K.M. Nutrient budgets and status in three pairs of conventional and alternative mixed cropping farms in Canterbury, New Zealand. Agr. Ecosyst. Environ. 1995, 52, 149–162.
21. Løes, A.-K.; Øgaard, A.F. Long-term changes in extractable soil phosphorus (P) in organic dairy farming systems. Plant Soil 2001, 237, 321–332.
22. Cornforth, I.S.; Sinclair, A.G. Model for calculating maintenance phosphate requirements for grazed pastures. New Zeal. J. Exp. Agr. 1982, 10, 53–61.
23. Gosling, P.; Shepherd, M. Long-term changes in soil fertility in organic arable farming systems in England, with particular reference to phosphorus and potassium. Agr. Ecosyst. Environ. 2005, 105, 425–432.
24. Soder, K.J.; Holden, L.A. Use of anionic salts with grazing prepartum dairy cows. Prof. Anim. Scientist 1999, 15, 278–285.
25. Bulluck, L.R.; Brosius, M.; Evanylo, G.K.; Ristaino, J.B. Organic and synthetic fertility amendments influence soil microbial, physical and chemical properties on organic and conventional farms. Appl. Soil Ecol. 2002, 19, 147–160.

26. National Academy of Sciences. In Watershed Management for Potable Water Supply: Assessing the New York City Strategy; National Academy Press: Washington, DC, USA, 2000.
27. Whitehead, D.C. In Nutrient Elements in Grassland: Soil-Plant-Animal Relationships; CABI: Wallingford, Oxon, UK, 2000.
28. Peterson, P.R.; Gerrish, J.R. Grazing management affects manure distribution by beef cattle. Available online: http://aes.missouri.edu/fsrc/research/afgc95pp.stm (accessed on 19 May 2013).
29. Cederberg, C.; Mattsson, B. Life cycle assessment of milk production–a comparison of conventional and organic farming. J. Clean. Prod. 2000, 8, 49–60.
30. Ryschawy, J.; Choisis, N.; Choisis, J.P.; Joannon, A.; Gibon, A. Mixed crop-livestock systems: An economic and environmental-friendly way of farming? Animal 2012, 6, 1722–1730.
31. Drinkwater, L.E.; Wagoner, P.; Sarrantonio, M. Legume-based cropping systems have reduced carbon and nitrogen losses. Nature 1998, 396, 262–265.
32. Peoples, M.B.; Baldock, J.A. Nitrogen dynamics of pastures: Nitrogen fixation inputs, the impact of legumes on soil nitrogen, fertility, and the contributions of fixed nitrogen to Australian farming systems. Aust. J. Exp. Agr. 2001, 41, 324–346.
33. Khan, S.A.; Mulvaney, R.L.; Ellsworth, T.R.; Boast, C.W. The myth of nitrogen fertilization for soil carbon sequestration. J. Environ. Qual. 2007, 36, 1821–1832.
34. Blanco-Canqui, H.; Schlegel, A.J. Implications of inorganic fertilization of irrigated corn on soil properties: Lessons learned after 50 years. J. Environ. Qual. 2013, 42, 861–871.
35. Brady, N. The Nature and Property of Soils, 10th ed.; Pearson: Upper Saddle River, NJ, USA, 1990.
36. James, D.W.; Hanks, R.J.; Jurinak, J.J. Modern Irrigated Soils; Wiley: New York, NY, USA, 1982.
37. Soder, K.J.; Stout, W.L. Effect of soil type and fertilization level on mineral concentration of pasture: Potential relationships to ruminant performance and health. J. Anim. Sci. 2003, 81, 1603–1610.
38. Ball, D.M.; Hoveland, C.S.; Lacefield, G.D. Southern Forages: Modern Concepts for Forage Crop Management, 4th ed.; Graphic Communications Corporation: Lawrenceville, GA, USA, 2007.
39. Church, D.C. The Ruminant Animal: Digestive Physiology and Nutrition; Waveland Press, Inc.: Prospect Heights, IL, USA, 1988.
40. Van Soest, P.J. Nutritional Ecology of the Ruminant, 2nd ed.; Cornell University Press: Ithaca, New York, NY, USA, 1994.
41. Goff, J.P.; Horst, R.L. Effects of the addition of potassium or sodium, but not calcium, to prepartum rations on milk fever in dairy cows. J. Dairy Sci. 1997, 80, 176–186.
42. Hardeng, F.; Edge, V.L. Mastitis, ketosis, and milk fever in 31 organic and 93 conventional Norwegian dairy herds. J. Dairy Sci. 2001, 84, 2673–2679.
43. Zollitsch, W.; Kristensen, T.; Krutzinna, C.; MacNaeihde, F.; Younie, D. Feeding for Health and Welfare: The Challenge of Formulating Well-balanced Rations in Organic Livestock Production. In Animal Health and Welfare in Organic Agriculture; Vaarst, M., Roderick, S., Lund, V., Lockeretz, W., Eds.; CABI Publishing: Wallingford, Oxon, UK, 2004; pp. 329–356.

44. Beal, S. Free Choice Smorgasbord Vitamin and Mineral Supplementation for Livestock. NODPA News, 2010, 10, 8–9, 30–31.
45. Weller, R.F.; Bowling, P.J. The yield and quality of plant species grown in mixed organic swards. In Organic Meat and Milk from Ruminants; Kyriazakis, I., Zervas, G., Eds.; Wageningen Academic Publishers: Wageningen, The Netherlands, 2002; pp. 177–183.
46. Soder, K.J.; Sanderson, M.A.; Stack, J.L.; Muller, L.D. Intake and performance of lactating cows grazing diverse forage mixtures. J. Dairy Sci. 2006, 89, 2158–2167.
47. Allen, V.G.; Pond, K.R.; Saker, K.E.; Fontenot, J.P.; Bagley, C.P.; Ivy, R.L.; Evans, R.R.; Schmidt, R.E.; Fike, J.H.; Zhang, X.; et al. Tasco: Influence of a brown seaweed on antioxidants in forages and livestock–A review. J. Anim. Sci. 2001, 79, E21–E31.
48. Berry, M.H.; Turk, K.L. The value of kelp meal in rations for dairy cattle. J. Dairy Sci. 1944, 27, 861–866.
49. Soule, G.M.; Brito, A.F.; Miranda, A.; Chase, L.; Whitehouse, N.L.; Fletcher, E.S.; Antaya, N.T. Effects of kelp meal on performance and structural growth of conventional and organic dairy calves. J. Dairy Sci. 2012, 95(Suppl. 2), 109.
50. Younie, D.; Thamsborg, S.M.; Ambrosini, F.; Roderick, S. Grassland Management and Parasite Control. In Animal Health and Welfare in Organic Agriculture; Vaarst, M., Roderick, S., Lund, V., Lockeretz, W., Eds.; CABI Publishing: Wallingford, Oxon, UK, 2004; pp. 308–328.
51. Hoste, H.; Jackson, F.; Athanasiadou, S.; Thamsborg, S.M.; Hoskin, S.O. The effects of tannin-rich plants on parasitic nematodes in ruminants. Trends Parasitol. 2006, 22, 253–261.
52. Bernard, G.; Worku, M.; Ahmedna, M. The effects of diatomaceous earth on parasite-infected goats. Bull. Georgian Natl. Acad. Sci. 2009, 3, 129–135.
53. Burke, J.M.; Wells, A.; Casey, P.; Miller, J.E. Garlic and papaya lack control over gastrointestinal nematodes in goats and lambs. Vet. Parasitol. 2009, 159, 171–174.
54. Kunkel, J.R.; Murphy, W.M.; Rogers, D.; Dugdale, D.T. Seasonal control of gastrointestinal parasites among dairy heifers. Bovine Pr. 1983, 18, 54–57.
55. Myers, G.H. Strategies to control internal parasites in cattle and swine. J. Anim. Sci. 1988, 66, 1555–1564.
56. Worthington, V. Nutritional quality of organic versus conventional fruits, vegetables, and grains. J. Altern. Complem. Med. 2001, 7, 161–173.
57. Heaton, S. Assessing organic food quality: Is it better for you? In UK Organic Research 2002, Proceedings of the COR Conference, 26–28 March 2002; Powell, J., Ed.; University of Wales: Aberystwyth, UK, 2002; pp. 55–60.
58. Bourn, D.; Prescott, J. A comparison of the nutritional value, sensory qualities, and food safety of organically and conventionally produced foods. CRC Cr. Rev. Food Sci. 2002, 42, 1–34.
59. Magkos, F.; Arvaniti, F.; Zampelas, A. Organic food: Nutritious food or food for thought? A review of the evidence. Int. J. Food Sci. Nutr. 2003, 54, 357–371.
60. Givens, D.I. Milk and meat in our diet: Good or bad for health? Animal 2010, 4, 1941–1952.
61. Shingfield, K.J.; Bonnet, M.; Scollan, N.D. Recent developments in altering the fatty acid composition of ruminant-derived foods. Animal 2013, 7, 132–162.

62. Lock, A.L.; Garnsworthy, P.C. Seasonal variation in milk conjugated linoleic acid and delta9-desaturase activity in dairy cows. Livest. Prod. Sci. 2003, 79, 47–59.
63. Palmquist, D.L.; Beaulieu, D.; Barbano, D.M. Feed and animal factors influencing milk fat and composition. J. Dairy Sci. 1993, 76, 1753–1771.
64. Chilliard, Y.; Ferlay, A. Dietary lipids and forages interactions on cow and goat milk fatty acid composition and sensory properties. Reprod. Nutr. Dev. 2004, 44, 467–492.
65. Chilliard, Y.; Glasser, F.; Ferlay, A.; Bernard, L.; Rouel, J.; Doreau, M. Diet, rumen biohydrogenation and nutritional quality of cow and goat milk fat. Eur. J. Lipid Sci. Tech. 2007, 109, 828–855.
66. Elgersma, A.; Tamminga, S.; Dijkstra, J. Lipids in herbage. In Fresh Herbage for Dairy Cattle; Elgersma, A., Dijikstra, J., Tamminga, S., Eds.; Springer: Wageningen, The Netherlands, 2006; pp. 175–194.
67. Ferlay, A.; Agabriel, C.; Sibra, C.; Journal, C.; Martin, B.; Chilliard, Y. Tanker milk variability in fatty acids according to farm feeding and husbandry practices in a French semi-mountain area. Dairy Sci. Technol. 2008, 88, 193–215.
68. Chilliard, Y.; Ferlay, A.; Doreau, M. Effect of different types of forages, animal fat or marine oils in cow's diet on milk fat secretion and composition, especially conjugated linoleic acid (CLA) and polyunsaturated fatty acids. Livest. Prod. Sci. 2001, 70, 31–48.
69. Dewhurst, R.J.; Shingfield, K.J.; Lee, M.R.F.; Scollan, N.D. Increasing the concentrations of beneficial polyunsaturated fatty acids in milk produced by dairy cows in high-forage systems. Anim. Feed Sci. Tech. 2006, 131, 168–206.
70. Elgersma, A.; Tamminga, S.; Ellen, G. Modifying milk composition through forage. Anim. Feed Sci. Tech. 2006, 131, 207–225.
71. Palupi, E.; Jayanegara, A.; Ploeger, A.; Kahl, J. Comparison of nutritional quality between conventional and organic dairy products: A meta-analysis. J. Sci. Food Agric. 2012, 92, 2774–2781.
72. Ip, C.; Banni, S.; Angioni, E.; Carta, G.; McGinley, J.; Thompson, H.J.; Barbano, D.; Bauman, D. Conjugated linoleic acid–enriched butter fat alters mammary gland morphogenesis and reduces cancer risk in rats. J. Nutr. 1999, 129, 2135–2142.
73. Butler, G.; Nielsen, J.H.; Slots, T.; Seal, C.; Eyre, M.D.; Sanderson, R.; Leifert, C. Fatty acid and fat-soluble antioxidant concentrations in milk from high- and low-input conventional and organic systems: Seasonal variation. J. Sci. Food Agric. 2008, 88, 1431–1441.
74. Butler, G.; Stergiadis, S.; Seal, C.; Eyre, M.; Leifert, C. Fat composition of organic and conventional retail milk in northeast England. J. Dairy Sci. 2011, 94, 24–36.
75. Wyss, U.; Miinger, A.; Collomb, M. Variation of fatty acid content in grass and milk during the grazing season. Grassland Sci. Eur. 2010, 15, 422–424.
76. Duckett, S.K.; Neel, J.P.; Sonon, R.N.; Fontenot, J.P.; Clapham, W.M.; Scaglia, G. Effects of winter stocker growth rate and finishing system on: II. Ninth-tenth-eleventh-rib composition, muscle color, and palatability. J. Anim. Sci. 2007, 85, 2691–2698.
77. Duckett, S.K.; Neel, J.P.S.; Lewis, R.M.; Fontenot, J.P.; Clapham, W.M. Effects of forage species or concentrate finishing on animal performance, carcass and meat quality. J. Anim. Sci. 2013, 91, 1454–1467.

78. Neel, J.P.; Fontenot, J.P.; Clapham, W.M.; Duckett, S.K.; Felton, E.E.; Scaglia, G.; Bryan, W.B. Effects of winter stocker growth rate and finishing system on: I. Animal performance and carcass characteristics. J. Anim. Sci. 2007, 85, 2012–2018.
79. Scaglia, G.; Fontenot, J.P.; Swecker, W.S.; Corl, B.A.; Duckett, S.K.; Boland, H.T.; Smith, R.; Abaye, A.O. Performance, carcass, and meat characteristics of beef steers finished on 2 different forages or on a high-concentrate diet. Prof. Anim. Sci. 2012, 28, 194–203.
80. Bennett, L.L.; Hammond, A.C.; Williams, M.J.; Kunkle, W.E.; Johnson, D.D.; Preston, R.L.; Miller, M.F. Performance, carcass yield, and carcass quality characteristics of steers finished on rhizoma peanut (Arachis glabrata)-tropical grass pasture or concentrate. J. Anim. Sci. 1995, 73, 1881–1887.
81. Hedrick, H.B.; Paterson, J.A.; Matches, A.G.; Thomas, J.D.; Morrow, R.E.; Stringer, W.C.; Lipsey, R.J. Carcass and palatability characteristics of beef produced on pasture, corn silage and corn grain. J. Anim. Sci. 1983, 57, 791–801.
82. Muir, P.D.; Deaker, J.M.; Bown, M.D. Effects of forage- and grain-based feeding systems on beef quality: A review. New Zeal. J Agr. Sci. 1998, 41, 623–635.
83. Scollan, N.; Hocquette, J.F.; Nuernberg, K.; Dannenberger, D.; Richardson, I.; Moloney, A. Innovations in beef production systems that enhance the nutritional and health value of beef lipids and their relationship with meat quality. Meat Sci. 2006, 74, 17–33.
84. Duckett, S.K.; Neel, J.P.; Fontenot, J.P.; Clapham, W.M. Effects of winter stocker growth rate and finishing system on: III. Tissue proximate, fatty acid, vitamin, and cholesterol content. J. Anim. Sci. 2009, 87, 2961–2970.
85. Noci, F.; Monahan, F.J.; French, P.; Moloney, A.P. The fatty acid composition of muscle fat and subcutaneous adipose tissue of pasture-fed beef heifers: Influence of the duration of grazing. J. Anim. Sci. 2005, 83, 1167–1178.
86. Gilmore, L.A.; Walzem, R.L.; Crouse, S.F.; Smith, D.R.; Adams, T.H.; Vaidyanathan, V.; Cao, X.; Smith, S.B. Consumption of high-oleic acid ground beef increases HDL-cholesterol concentration but both high- and low-oleic acid ground beef decrease HDL particle diameter in normocholesterolemic men. J. Nutr. 2011, 141, 1188–1194.
87. Kris-Etherton, P.M.; Pearson, T.A.; Wan, Y.; Hargrove, R.L.; Moriarty, K.; Fishell, V.; Etherton, T.D. High–monounsaturated fatty acid diets lower both plasma cholesterol and triacylglycerol concentrations. Am. J. Clin. Nutr. 1999, 70, 1009–1015.
88. Chait, A.; Brazg, R.L.; Tribble, D.L.; Krauss, R.M. Susceptibility of small, dense, low-density lipoproteins to oxidative modification in subjects with the atherogenic lipoprotein phenotype, pattern B. Am. J. Med. 1993, 94, 350–356.
89. Kouba, M. Quality of organic animal products. Livest. Prod. Sci. 2003, 80, 33–40.
90. Caroprese, M.; Marzano, A.; Marino, R.; Gliatta, G.; Muscio, A.; Sevi, A. Flaxseed supplementation improves fatty acid profile of cow milk. J. Dairy Sci. 2010, 93, 2580–2588.
91. Kronberg, S.L.; Scholljegerdes, E.J.; Leper, A.N.; Berg, E.P. The effect of flaxseed supplementation on growth, carcass characteristics, fatty acid profile, retail shelf life, and sensory characteristics of beef from steers finished on grasslands of the northern Great Plains. J. Anim. Sci. 2011, 89, 2892–2903.

92. Donovan, D.C.; Schingoethe, D.J.; Baer, R.J.; Ryali, J.; Hippen, A.R.; Franklin, S.T. Influence of dietary fish oil on conjugated linoleic acid and other fatty adids in milk fat from lactating dairy cows. J. Dairy Sci. 2000, 83, 2620–2628.
93. Olesen, J.E.; Schelde, K.; Weiske, A.; Weisbjerg, M.R.; Asman, W.A.H.; Djurhuus, J. Modelling greenhouse gas emissions from European conventional and organic dairy farms. Agr. Ecosyst. Environ. 2006, 112, 207–220.
94. Thomassen, M.A.; van Calker, K.J.; Smits, M.C.J.; Iepema, G.L.; de Boer, I.J.M. Life cycle assessment of conventional and organic milk production in the Netherlands. Agr. Syst. 2008, 96, 95–107.
95. Lynch, D.; MacRae, R.; Martin, R. The carbon and global warming potential impacts of organic farming: Does it have a significant role in an energy constrained world? Sustainability 2011, 3, 322–362.
96. Capper, J.L.; Castaneda-Gutierrez, E.; Cady, R.A.; Bauman, D.E. The environmental impact of recombinant bovine somatotropin (rbST) use in dairy production. Proc. Natl. Acad. Sci. USA 2008, 105, 9668–9673.
97. FAO. Greenhouse Gas Emissions from the Dairy Sector: A Life Cycle Assessment; Food and Agiculture Organization of the United Nations, Animal Production and Health Division: Rome, Italy, 2010.
98. United States Department of Agriculture, Economic Research Service (USDA-ERS). Farm Milk Production. Available online: http://www.ers.usda.gov/topics/animal-products/dairy/background.aspx (accessed on 24 March 2013).
99. Schultz, M. Organic Dairy Profile; Agricultural Marketing Resource Center: Washington, DC, USA, 2013.
100. 100. Peters, G.M.; Rowley, H.V.; Wiedemann, S.; Tucker, R.; Short, M.D.; Schulz, M. Red meat production in Australia: Life cycle assessment and comparison with overseas studies. Envir. Sci. Tech. 2010, 44, 1327–1332.
101. Crosson, P.; Shalloo, L.; O'Brien, D.; Lanigan, G.J.; Foley, P.A.; Boland, T.M.; Kenny, D.A. A review of whole farm systems models of greenhouse gas emissions from beef and dairy cattle production systems. Anim. Feed Sci. Tech. 2011, 166–167, 29–45.
102. Kim, S.; Dale, B.E. Cumulative energy and global warming impact from the production of biomass for biobased products. J. Ind. Ecol. 2004, 7, 147–162.
103. CAST. Carbon Sequestration and Greenhouse Gas Fluxes in Agriculture: Challenges and Opportunities; Task Force Report No. 142; Council for Agricultural Science and Technology: Ames, IA, USA, 2011.
104. Paustian, K.; Antle, J.M.; Sheehan, J.; Paul, E.A. Agriculture's Role in Greenhouse Gas Mitigation; Pew Center on Global Climate Change: Arlington, VA, USA, 2006.
105. Follett, R.F.; Reed, D.A. Soil carbon sequestration in grazing lands: Societal benefits and policy implications. Rangel Ecol. Manag. 2010, 63, 4–15.
106. Sayre, N.F.; Carlisle, L.; Huntsinger, L.; Fisher, G.; Shattuck, A. The role of rangelands in diversified farming systems: Innovations, obstacles, and opportunities in the USA. Ecol. Soc. 2012, 17, 43.
107. Greene, C. Data Track the Expansion of International and U.S. Organic Farming. Amber Waves 2007, 5, 36–37.

108. White, S.L.; Benson, G.A.; Washburn, S.P.; Green, J.T. Milk production and economic measures in confinement or pasture systems using seasonally calved Holstein and Jersey cows. J. Dairy Sci. 2002, 85, 95–104.
109. Dartt, B.A.; Lloyd, J.W.; Radke, B.R.; Black, J.R.; Baneene, J.B. A comparison of profitability and economic efficiencies between management-intensive grazing and conventionally managed dairies in Michigan. J. Dairy Sci. 1999, 82, 2412–2420.
110. Hanson, G.D.; Cunningham, L.C.; Morehart, M.J.; Parsons, R.L. Profitability of moderate intensive grazing of dairy cows in the Northeast. J. Dairy Sci. 1998, 81, 821–829.
111. Hanson, J.C.; Johnson, D.M.; Lichtenberg, E.; Minegishi, K. Competitiveness of management-intensive grazing dairies in the mid-Atlantic region from 1995 to 2009. J. Dairy Sci. 2013, 96, 1894–1904.
112. Parker, W.J.; Muller, L.D.; Buckmaster, D.R. Management and economic implications of intensive grazing on dairy farms in the northeastern states. J. Dairy Sci. 1992, 75, 2587–2597.
113. McBride, W.D.; Greene, C. Characteristics, Costs, and Issues for Organic Dairy Farming; Research Report Number 82; United States Department of Agriculture, Economic: Washington, DC, USA, 2009.
114. Berentsen, P.B.; Kovacs, K.; van Asseldonk, M.A. Comparing risk in conventional and organic dairy farming in the Netherlands: an empirical analysis. J. Dairy Sci. 2012, 95, 3803–3811.
115. Lund, V. Natural living—a precondition for animal welfare in organic farming. Livest. Sci. 2006, 100, 71–83.
116. Bocquier, F.; Gonzalez-Garcia, E. Sustainability of ruminant agriculture in the new context: Feeding strategies and features of animal adaptability into the necessary holistic approach. Animal 2010, 4, 1258–1273.
117. Hodgson, J. Grazing Management: Science into Practice; Longman Scientific and Technical: Harlow, Essex, UK, 1990.
118. Dillon, P.; Hennessy, T.; Shalloo, L.; Thorne, F.; Horan, B. Future outlook for the Irish dairy industry: A study of international competitiveness, influence of international trade reform and requirement for change. Int. J. Dairy Technol. 2008, 61, 16–29.
119. Peyraud, J.L.; Delagarde, R. Managing variations in dairy cow nutrient supply under grazing. Animal 2013, 7, 57–67.
120. Butler, L.J. The economics of organic milk production in California: A comparison with conventional costs. Am. J. Alternative Agr. 2002, 17, 83–91.
121. Rotz, C.A.; Kamphuis, G.H.; Karsten, H.D.; Weaver, R.D. Organic dairy production systems in Pennsylvania: A case study evaluation. J. Dairy Sci. 2007, 90, 3961–3979.
122. Rouquette, F.M.; Redmon, L.A.; Aiken, G.E.; Hill, G.M.; Sollenberger, L.E.; Andrae, J. ASAS Centennial Paper: Future needs of research and extension in forage utilization. J. Anim. Sci. 2009, 87, 438–446.
123. Younie, D. An organic approach to meat production. Food Sci. Technol. Today 1992, 6, 163–166.
124. Fernández, M.I.; Woodward, B.W. Comparison of conventional and organic beef production systems I. Feedlot performance and production costs. Livest. Prod. Sci. 1999, 61, 213–223.

125. Lewis, J.M.; Klopfenstein, T.J.; Pfeiffer, G.A.; Stock, R.A. An economic evaluation of the differences between intensive and extensive beef production systems. J. Anim. Sci. 1990, 68, 2506–2516.
126. Reichl, R. Michael Pollan and Ruth Reichl hash out the food revolution. Smithsonian 2013, 44, 74–83.
127. Papendick, R.I.; Elliot, L.F.; Dalgren, R.B. Environmental consequences of modern production agriculture: How can alternative agriculture address these concerns? Am. J. Alternative Agr. 1986, 1, 3–10.
128. Lasley, P.; Hoiberg, E.; Bultena, G. Is sustainable agriculture an elixir for rural communities? Am. J. Alternative Agr. 1993, 8, 133–139.
129. Realini, C.E.; Duckett, S.K.; Brito, G.W.; Dalla Rizza, M.; de Mattos, D. Effect of pasture vs. concentrate feeding with or without antioxidants on carcass characteristics, fatty acid composition, and quality of Uruguayan beef. Meat Sci. 2004, 66, 567–577.

CHAPTER 5

CHANGES OF SOIL BACTERIAL DIVERSITY AS A CONSEQUENCE OF AGRICULTURAL LAND USE IN A SEMI-ARID ECOSYSTEM

GUO-CHUN DING, YVETTE M. PICENO, HOLGER HEUER, NICOLE WEINERT, ANJA B. DOHRMANN, ANGEL CARRILLO, GARY L. ANDERSEN, THELMA CASTELLANOS, CHRISTOPH C. TEBBE, AND KORNELIA SMALLA

5.1 INTRODUCTION

Converting natural land into arable soils results in losses to the landscapes characterized by a typical indigenous flora and fauna. Frequently, terrestrial ecosystem diversity is being reduced by replacing indigenous flora with a few crops. The ecological consequences of such transitions have been addressed in several studies focusing on land degradation [1], [2], [3], losses of macro-biodiversity [4], [5], nutrient exhaustion in soils [3], sustainability [6], [7] and restoration [8]. Soil microorganisms, including protozoa, fungi, bacteria and archaea, are essential for the proper functioning and sustainability of ecosystems [9], [10], [11]. Moreover, a high microbial diversity is assumed to be critical for the stability of ecosystems

This chapter was originally published under the Creative Commons Attribution License. Ding G-C, Piceno YM, Heuer H, Weinert N, Dohrmann AB, Carrillo A, Andersen GL, Castellanos T, Tebbe CC, and Smalla K. Changes of Soil Bacterial Diversity as a Consequence of Agricultural Land Use in a Semi-Arid Ecosystem. PLoS ONE ***8****,3 (2013). doi:10.1371/journal.pone.0059497.*

by providing functional diversity and redundancy [12], [13]. Changes in vegetation as well as intensive agricultural practices were shown to affect soil microbial community composition and activity [14], [15] and soil physicochemical properties [3]. The influence of land use and management on soil microorganisms was addressed in several recent studies [16], [17]. However, the information acquired is still not sufficient as a systematic identification of taxa responding to the transition in land use was not done.The studies investigated soils from various geographic sites in Australia, The Netherlands, and Brazil [16] although a comparison of the results might also be difficult due to the differences in the experimental designs and the resolution level of the methods used.

The Santo Domingo Valley is an agricultural area within the Southern Sonoran Desert, Baja California Sur, Mexico, which is entirely dependent on irrigation water collected from wells. Most of the farmland has been developed during the 1950's and 60′s with cotton and wheat as the main crops [18]. Since then, other crops have been cultivated, e.g., oat, sorghum, chickpea, maize, and alfalfa. Due to declining water availability and increasing problems with soil salinization, yields have decreased since 1991 [18]. Most of the arable land is bordered by natural sarcocaulescent scrubland, including different crassicaulent plants, succulent cacti, woody trees and shrubs [19]. Some of these areas are utilized for grazing goats [20], [21] and thus might be impacted by fecal depositions. The soils in the Santo Domingo Valley belong to the hyposodic calcisols, which are typical of many semi-arid ecosystems. The two sites selected provided almost ideal conditions to study the effects of agricultural land use on the microbial communities in these ecosystems as agricultural fields were in direct vicinity to the scrubland and the soils were typical of semi-arid deserts.

In the present study, bacterial soil communities from arable fields with alfalfa and the adjacent scrubland at two sites 50 km apart from each other were compared by denaturing gradient gel electrophoresis (DGGE) and PhyloChip analysis of16S rRNA gene fragments amplified from total community DNA to evaluate the influence of land use. Both methods are complementary but have clear differences. DGGE provides information on the relative abundance of all amplified dominant populations and thus is more suitable for comparative analysis of the community composition. In the present study the bacterial community analyses by DGGE was performed at different taxonomic levels, *Bacteria, Actinobacteria, Alpha-* and *Betaproteobacteria* in order to analyze not only dominant bacteria. The so-called PhyloChip developed by Brodie et al. [36] offers the potential to detect 8741 operational taxonomic units (OTUs) and the dataset is ideal for identifying taxa containing a high proportion of OTUs with treatment dependent significantly increased or decreased abundance. Therefore, the PhyloChip dataset was used to analyze whether, and if so which bacterial taxa responded to agricultural use, and multivariate statistics was applied to explore the relationship between discriminative soil parameters and responsive taxonomic groups.

5.2 MATERIALS AND METHODS

5.2.1 SITE AND SAMPLING

Two typical sites in the Santo Domingo valley, Baja California, Mexico (site 1: N25°06′31.5″ W111°32′34.3″, altitude 70; site 2: N25°16′5.2″ W111° 36′ 02.5″, altitude 90; the distance between site 1 and site 2 is ca 50 km; No specific permits were required for the described field studies. The location is not protected in any way. The field studies did not involve endangered or protected species.) were selected. Soil samples were taken on 30 April 2007 from two covers, i.e., a field planted with alfalfa (*Medicago sativa*) and the adjacent natural scrubland. Four replicates per site and cover were taken respectively from plots (1 m×1 m squares) that were at least 20 m apart from each other. After removing the top 1–2 cm soil layer, the soil from 2–15 cm depth was mixed with a shovel and approx. 2 kg of soil was sampled, put into plastic bags and transferred to the laboratory. Within 24 h after sampling, the soils were sieved through a 2 mm mesh and aliquots were used for microbiological and chemical analysis.

5.2.2 SOIL PROPERTIES

All soils were analyzed by the certified laboratory at CIBNOR (La Paz, Mexico). Briefly, soil particle size determinations were conducted with the sedimentation method [22], pH was determined in 1:1 (wt/vol) diluted water suspensions [23], electrical conductivity was determined with a CO150 conductivity meter according to the manufacturer's instructions (Hach Company, Loveland, CO, USA). Total organic matter was measured using the reduction of potassium dichromate method of Walkley and Black, as described by Nelson and Sommers [24]. Furthermore, ammonium (NH_4^+), nitrite and nitrate [25], calcium (Ca^{2+}) [26], magnesium (Mg^{2+}), phosphate, sulfate [27] were quantified. All soil parameters measured are given in Table 1.

TABLE 1: Physico-chemical characteristics (average ± standard deviation) for soils from different sites and land use.

	Site 1		Site 2	
	Alfalfa	scrubland	Alfalfa	scrubland
pH value	8.65 ± 0.12	8.7 ± 0.27	8.62 ± 0.06	8.1 ± 0.64
Electric conductivity [mS cm^{-1}]	0.86 ± 0.14	0.59 ± 0.18	1.45 ± 0.33	1.3 ± 1.71
CA^{2+} [mg kg^{-1}]	8.67 ± 2.16	9.89 ± 3.36	10.83 ± 4.56	13.2 ± 14.17
Mg^{2+} [mg kg^{-1}]	5.42 ± 1.38	5 ± 0.98	13.34 ± 5.32	6.63 ± 6.46
K^+ [mg kg^{-1}]	4.59 ± 1.62	4.8 ± 1.99	5.09 ± 2.7	4.01 ± 2.14
Na^+ [mg kg^{-1}]* +	33.22 ± 6.34	13.86 ± 9.37	44.56 ± 8.2	26.4 ± 32.55
Phosphate [mg kg^{-1}]*** –	5.2 ± 1.33	41.62 ± 19.15	7.21 ± 3.32	30.31 ± 17.97
NH_4^+-N [mg kg^{-1}]	10.14 ± 4.1	7.44 ± 1.53	10.48 ± 2.17	10.14 ± 3.3
Total nitrogen [mg kg^{-1}]	482.39 ± 179.03	450.48 ± 76.27	358.29 ± 130.19	308.65 ± 170.74
Cl^- [mg kg^{-1}]	37.6 ± 5.28	11.57 ± 7.57	60.23 ± 15.41	65.64 ± 110.05
NO_2^-N [mg kg^{-1}]	0.05 ± 0.03	0.03 ± 0.03	0.09 ± 0.16	0.03 ± 0.04
NO_3^-N [mg kg^{-1}]*+	2.62 ± 0.72	0.66 ± 0.5	1.18 ± 1.92	0.23 ± 0.22
Sulphate [mg kg^{-1}]***+	8.35 ± 1.63	4.79 ± 1.33	29.28 ± 6.24	2.97 ± 0.92
Organic matter [% volumetric]*–	0.45 ± 0.06	0.79 ± 0.18	0.44 ± 0.11	0.58 ± 0.32

*Two-way ANOVA was used to test for significant differences in soil parameters between land use. The significant level between land use was indicated as *$p < 0.05$; ***$p < 0.001$; +: significantly increased in agricultural soils; –: significantly decreased in agricultural soils.*

5.2.3 TOTAL COMMUNITY (TC) DNA

TC DNA was extracted from 0.5 g of soil after a harsh lysis step (FastPrep FP120 bead beating system, MP Biomedicals, Santa Ana, Carlsbad, CA, USA) by means of the BIO-101 DNA spin kit for soil (Q-Biogene). The DNA was purified using the Geneclean Spin Kit (Q-Biogene) according to the manufacturer's instructions. Purified TC DNA was stored at −20°C.

5.2.4 PCR AMPLIFICATION OF 16S RRNA GENE FRAGMENTS AND DGGE ANALYSIS

Primer sets and PCR conditions employed in this study to amplify *Bacteria* [28], *Actinobacteria* [28], and *Alpha-* and *Betaproteobacteria* [29], [30] and relevant information is provided in Table S1. DGGE of the 16S rRNA gene amplicons was performed according to Gomes et al. [31]. The gel was silver-stained according to Heuer et al. [32]. DGGE profiles were analyzed by GelCompar 4.5. Dendrograms were constructed by means of unweighted pair group method using arithmetic averages (UPGMA) based on pairwise Pearson correlation indices, which were also subjected to permutation tests with a modified version of PERMTEST software [33]. Box-Whisker plots were generated using R (http://www.R-project.org) based on dissimilarities (1- Pearson correlation indices) between samples within the same treatment.

5.2.5 PCR AMPLIFICATION OF 16S RRNA GENES AND PHYLOCHIP ANALYSIS

TC DNA extracts from three replicates per site and cover were amplified using universal 16S rRNA gene primers (27f 5′- AGAGTTTGATCCTGGCTCAG-3′; 1492r 5′- GGTTACCTTGTTACGACTT-3′) and an 8-temperature gradient PCR. At each temperature, approximately 5 ng of TC DNA was used in 25 μl reactions (final concentrations were 1× Ex Taq Buffer with 2 mM $MgCl_2$, 300 nM each primer (27 f and 1492 r), 200 μM each dNTP (TaKaRa), 25 μg bovine serum albumin (Roche Applied

Science, Indianapolis, IN, USA), and 0.625 U Ex Taq (TaKaRa Bio, Inc., through Fisher Scientific, Pittsburg, PA, USA)). The amplifications were performed with an iCycler (Bio-Rad, Hercules, CA, USA) as previously described by Weinert et al. [34]. PCR products from each annealing temperature (48–58°C) for each sample were combined, concentrated, quantified, and an amount of 500 ng product was applied to each G2 PhyloChip (Second Genome Inc., San Bruno, CA, USA) following previously described procedures [35]. The PhyloChip used in the present study contained approximately 500,000 probes (25-mer oligos) targeting 8,364 bacterial operational taxonomic units (OTUs). An OTU was considered present if more than 90 percent of the probe pairs representing this OTU showed a positive hybridization signals [36].

5.2.6 STATISTICAL ANALYSIS

An OTU-level report was produced mainly according to Brodie et al. [36] (background subtraction, detection and quantification criteria) except for the addition of normalizing array data by the average total array signal intensity [37]. The signal intensities of OTUs called absent were shifted to 1 to avoid errors in subsequent log transformation. Statistical analyses were done with the software package R 2.14 (http://www.r-project.org/).

Discriminative OTUs between the two different land uses were identified by multiple two-way ANOVA of log10-transformed signal intensities for each OTU (unadjusted $p<0.05$). Groups with a high proportion of OTUs significantly responding to land use were summarized at different taxonomic levels. The influence of land use on the community compositions of different taxonomic groups (from domain to family) was analyzed by a modified test based on five principal components [38]. Principal components analysis (PCA) was performed according to Weinert et al. [34] using adjusted (log10- transformed, centered, and standardized) signal intensities of all OTUs belonging to each taxonomic group.

Soil parameters responding to land use were also identified by multiple two-way ANOVA (unadjusted $p<0.05$). Heatmap analysis was performed based on the adjusted (centered and standardized) values for different soil parameters. To analyze the influence of these discriminative soil parameters on community composition of total bacteria or responsive phyla, redundancy analysis (RDA) was performed using the R add-on package 'vegan'. A forward selection of soil parameters was applied to avoid using collinear soil parameters in the same constrained ordination model. Only those parameters contributing significantly ($p<0.05$ via 1000 times permutation tests) to community variation were added to the model.

5.3 RESULTS

5.3.1 SIGNIFICANT EFFECTS OF LAND USE ON THE BACTERIAL COMMUNITY COMPOSITION REVEALED BY DGGE FINGERPRINTS

To compare the bacterial community in soils under different land use, 16S rRNA gene fragments of bacteria, *Alphaproteobacteria, Betaproteobacteria* and *Actinobacteria* PCR-products amplified from total community DNA of alfalfa or adjacent scrubland soils sampled at two sites were analyzed by DGGE (Figures S1, S2, S3, S4).

Pairwise Pearson correlation indices were subjected to permutation tests to determine the significance of the land use effects on the bacterial community structure. A significant effect of land use was found for bacteria and all bacterial groups analyzed for both sites (Table 2). The extent of the influence was dependent on the phylogenetic group analyzed (Table 2). The strongest influence of land use was found for the *Betaproteobacteria,* especially at site 1 (Table 2). In the community profile for *Betaproteobacteria*, a strong band was observed only in all replicates in soils with alfalfa from site 1 (Figure S4). For *Alphaproteobacteria,* the dissimilarities of bacterial community fingerprints of soil under different land use were comparable between both sites. The lowest yet still significant effect of land use was observed for *Actinobacteria* (Table 2). Significant differences in community composition between the two study sites were found mainly for alfalfa growing soils as opposed to scrubland soils (Table 2). Compared to the effects of land use, the influence of different sites on community composition was smaller except for *Betaproteobacteria* (Table 2), probably still due to the strong bands for alfalfa soils from site 1 (Figure S4). For bacteria including all bacterial subgroups analyzed, Box-Whisker plots revealed that the variability of the bacterial community compositions among replicates was generally lower for alfalfa soils than the scrubland soil, except for *Betaproteobacteria* at site 2 (Figure 1). The lowest variability was found for alfalfa soils from site 1 (Figure 1). In conclusion, a significant and taxonomic group-dependent effect of land use was observed for all four targeted phylogenetic groups. Variation in community composition for soils under arable farmland use generally was lower than that from scrubland sites.

TABLE 2: Percent dissimilarity of microbial DGGE fingerprints of different taxa for soils compared between alfalfa and scrubland or between site 1 and site 2.

		Bacteria	*Alpha-proteobacteria*	*Beta-proteobacteria*	*Actinobacteria*
Land use	Site 1	25*	20.3*	68.6*	2.7*
	Site 2	13.8*	25*	31.3*	3.7*
Site	Alfalfa	10.3*	11.8*	61.5*	3.7*
	Scrubland	3.8	4.9	5.8	2.8*

**Significant ($P < 0.05$) difference in community fingerprints between treatments as revealed by permutation tests.*

5.3.2 TAXA RESPONSIVE TO LAND USE DETERMINED BY PHYLOCHIP ANALYSIS

PhyloChip hybridizations were used to detect bacterial taxa with significant responses to land use. A total of 2,243 OTUs belonging to 44 phyla was detected (Table S2). The bacterial richness in terms of the number of detected OTUs was significantly ($p = 0.05$) higher for the arable field soils than for the scrubland soils.

OTUs responding to land use were identified by multiple two-way ANOVA and only taxa with a high proportion (>18% which is much higher than the unadjusted p-value) of discriminative OTUs were summarized in Table 3.

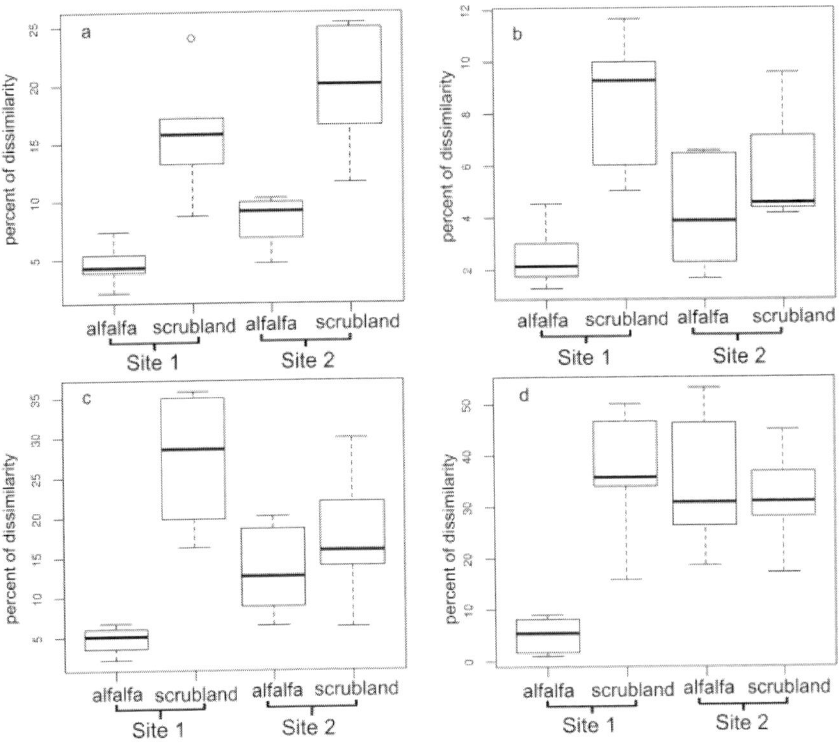

FIGURE 1: Boxplots of the variation of community structure under same sites and land use based on DGGE profiles for a: *Bacteria*; b: *Actinobacteria*; c: *Alphaproteobacteria*; d: *Betaproteobacteria*.

TABLE 3: Taxa and numbers (percent of all detected OTU belonging to each taxon) of OTUs significantly (unadjusted P<0.05) enriched in alfalfa or scrubland soil as identified by two-way ANOVA based on PhyloChip.

Phylum	Class	Order	Family	Alfalfa	Scrubland
Proteobacteria	*Alphaproteobacteria*	*Rhizobiales*	*Rhizobiaceae*	14 (77.8%)	0 (0.0%)
			Phyllobacteriaceae	9 (60.0%)	0 (0.0%)
		Sphingomondales		9 (18.0%)	1 (2.0%)
		Rhodobacteriales		20 (40.0%)	10 (20.0%)
	Betaproteobacteria	*Burkholderiales*	*Comamonadaceae*	36 (62.1%)	0 (0.0%)
	Gammaproteobacteria	*Alteromonadales*		15 (26.3%)	2 (3.5%)
		Pseudomonadales	*Pseudomonadaceae*	21 (53.8%)	1 (2.6%)
		Legionellales		1 (9.1%)	6 (54.5%)
	Deltaproteobacteria	*Desulfobacterales*		0 (0.0%)	19 (63.3%)
		Desulfovibrionales		0 (0.0%)	8 (50.0%)
		Syntrophobacterales		0 (0.0%)	6 (42.9%)
Firmicutes	*Clostridia*			6 (3.4%)	51 (28.8%)
Actinobacteria	***Actinobacteria***	*Acidimicrobiales*		1 (5.6%)	6 (33.3%)
		Actinomycetales	*Microbacteriaceae*	15 (65.2%)	0 (0.0%)
			Micromonosporaceae	6 (27.3%)	2 (9.1%)
			Mycobacteriaceae	0 (0.0%)	16 (76.2%)
			Micrococcaceae	13 (76.5%)	0 (0.0%)
			Nocardiaceae	2 (12.5%)	1 (6.3%)
			Pseudonocardiaceae	1 (8.3%)	4 (33.3%)
			Cellulomonadaceae	7 (63.6%)	0 (0.0%)

TABLE 3: *Cont.*

Phylum	Class	Order	Family	Alfalfa	Scrubland
		Rubrobacterales		0 (0.0%)	10 (50.0%)
Acidobacteria	*Acidobacteria*	*Acidobacteriales*	*Acidobacteria-ceae*	0 (0.0%)	21 (56.8%)
Bacteriodetes				22 (18.2%)	10 (8.3%)
Chloroflexi				1 (2.1%)	25 (53.2%)
Spirochaetes				0 (0.0%)	12 (30.0%)
Verrucomi-crobia				0 (0.0%)	13 (48.1%)
Gemmatimon-adetes				0 (0.0%)	6 (66.7%)
Bacteria (total)				295 (13.2%)	402 (17.9%)

All listed taxa have significantly different community structure between land uses except for Nocardiaceae. *Bold text: Taxa with significantly different community structure between sites.*

Compared to the soil from the scrublands, 13% of the OTUs (295 OTUs) were more abundant in soil from alfalfa fields, while more OTUs (402 accounting for 18%) were less abundant (Table 3), suggesting few taxa probably enriched in the alfalfa fields. A large proportion (more than 25%) of the OTUs belonging to the phyla of *Acidobacteria, Chloroflexi, Spirochaetes, Verrucomicrobia* and *Gemmatimonadetes* were significantly less abundant in alfalfa soils (Table 3), suggesting that the transition of scrubland soil into arable soils caused severe effects on the abundance of OTUs affiliated to these phyla. A negative effect also was observed for the class of *Clostridia* (*Firmicutes*), of which 29% of the OTUs detected had significantly lower signal intensities in the alfalfa soils (Table 3). The influence of land use on the phyla *Proteobacteria* and *Actinobacteria* was more complex as more than 40% of the OTUs belonging to three orders of the *Deltaproteobacteria* (*Desulfobacterales, Desulfovibrionales* and *Syntrophobacterales*) were less abundant while many taxa belonging to

Alphaproteobacteria (*Rhizobiales: Rhizobiaceae, Phyllobacteriaceae; Sphingomonadales; Rhodobacterales*), *Betaproteobacteria* (*Burkholderiales; Comamonadaceae*) and *Gammaproteobacteria* (*Alteromonadales; Pseudomonadales: Pseudomonadaceae*) contained a high proportion of OTUs (>18%) that were more abundant in alfalfa soils (Table 3). An exception for *Gammaproteobacteria* was the order of *Legionellales,* of which ca. 56% of the OTUs were more abundant in the scrubland soils (Table 3). Interestingly, for *Rhodobacterales* (*Alphaproteobacteria*) 40% of the OTUs were more abundant in alfalfa soils, while 20% of the OTUs were lower compared to scrubland soils (Table 3). A high proportion (>33%) of the OTUs belonging to two orders of *Actinobacteria* (*Acidimicrobiales* and *Rubrobacterales*) had lower relative abundance in natural scrublands compared to alfalfa soils. Several families in the order of *Actinomycetales* (*Microbacteriaceae, Micromonosporaceae, Micrococcaceae,* and *Cellulomonadaceae*) had more OTUs with significantly higher abundance in alfalfa soils than in the scrubland soils (Table 3). In contrast to these families, *Mycobacteriaceae* and *Pseudonocardiaceae* decreased in relative abundance with change in land use (Table 3).

A modified test based on the first five principal components [38] was applied to study the effect of land use on bacterial community composition of the taxonomic groups listed in Table 3. It revealed and confirmed dramatic differences in the abundance of specific bacterial community members in response to land use. All the taxonomic groups listed in Table 3 had significantly different community compositions, except for *Nocardiaceae* ($p = 0.07$) (Table 3). As already observed with the DGGE analyses, the variation in the bacterial community composition was higher in the soils from scrubland than from alfalfa fields (Figure S5). Compared to the land use, the two different locations had less influence on the community composition. Effects of the site were only detected for a few taxonomic groups such as the *Sphingomonadales, Rhodobacterales, Rubrobacterales* and *Actinomycetales*. These orders, however, were significantly different in their community composition between the two sites (Table 3).

5.3.3 CORRELATION OF LAND USE RESPONSIVE TAXA WITH LAND USE DEPENDENT SOIL PARAMETERS

Two-way ANOVA of soil physicochemical parameters revealed that the transition of scrubland into arable land resulted in a significantly increased concentration of sulphate, sodium, salinity, and the ratio of nitrate to total-N. The phosphate concentration and organic matter content were significantly lower in agricultural soils (Table 1).

Redundancy analysis was performed to find the correlation between discriminative physicochemical characteristics and variation in the community composition of phyla identified by PhyloChips with significant response to land use. Sulphate and phosphate (phosphate positively collinear with organic matter content) were the main factors that jointly influenced the bacterial community composition (Figure 2). In total 33% of the variation of the bacterial community could be significantly explained by sulphate and phosphate concentrations. Each factor (sulphate, phosphate and organic matter) independently could explain 17% to 21% of the variation. Sulphate and phosphate also explained a considerable amount of the variation within sub-communities of *Proteobacteria* (33%), to which most detected OTUs were affiliated (data not shown) (Figure S6). Discriminative soil parameters also explained a large amount of the variation within *Firmicutes* (19% variation explained by sulphate; Figure S7), *Actinobacteria* (40% explained by phosphate and sulphate; Figure S8), *Acidobacteria* (30% explained by nitrate; Figure S9), *Bacteroidetes* (16% explained by nitrate; Figure S10), *Chloroflexi* (30% explained by nitrate; Figure S11), *Verrucomicrobia* (24% explained by phosphate; Figure S12) and *Gemmatimonadetes* (49% explained by sulphate and sodium; Figure S13). None of these discriminative soil parameters explained the variation within *Spirochaetes*, which appeared to be linked to the pH values of the soils (explaining 32% variation; Figure S14). In general, a considerable amount of variation in the community composition of phyla responding to land use changes could be significantly explained by discriminative soil parameters, suggesting that these soil parameters and the bacterial community structure are strongly connected.

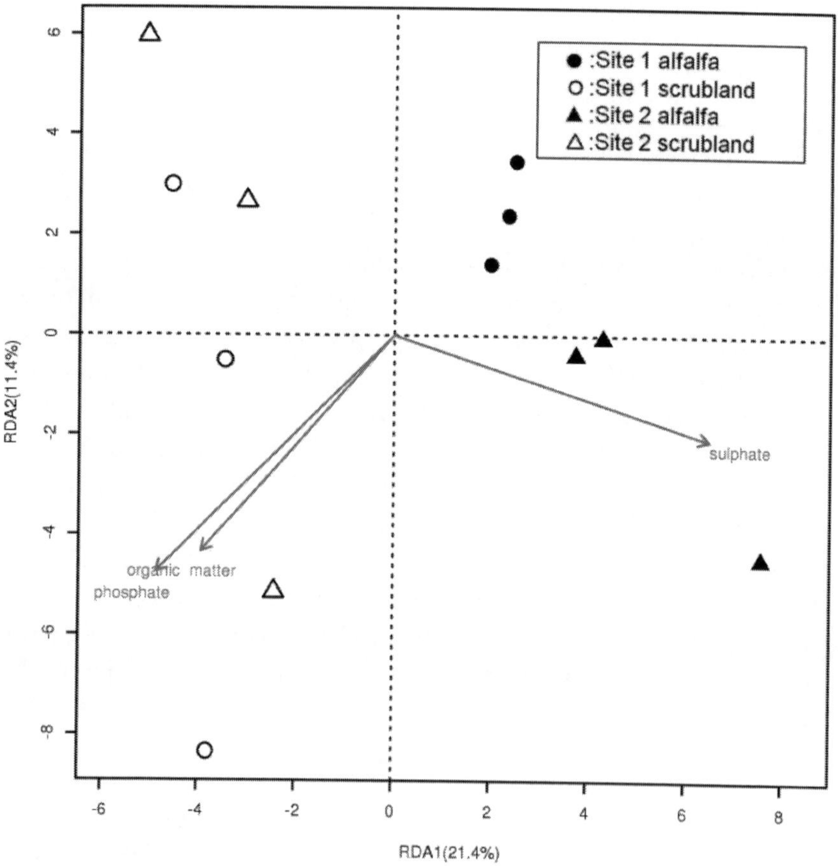

FIGURE 2: Redundancy analysis of the effect of discriminative soil parameters on bacterial communities using the PhyloChip data. Numbers in brackets indicate the percent of the total variance explained by each axis. Only these soil parameters which significantly ($p<0.05$ by 1000 times permutation tests) explained the bacterial community variation was shown.

5.4 DISCUSSION

Both DGGE and PhyloChip analysis revealed that the community composition of bacteria at various taxonomic levels differed significantly between the arable soils and the adjacent scrubland soils. The methods used to analyze the effects of the transition in land use were complementary but have clear differences which prevented the use of similar statistical analysis methods to analyze the datasets. In contrast to the DGGE fingerprints that are assumed to reflect the relative abundance of the dominant bacterial populations, the fluorescent signals detected after PhyloChip hybridizations do not necessarily reflect the relative abundance of OTUs. The strength of the PhyloChip approach is that the fluorescence signals of each OTU can be compared horizontally between treatments and thus allows to identify taxa with a high proportion of OTUs with treatment dependent changes in abundance. However, both methods used in this study to analyze PCR-amplified 16S rRNA gene fragments lead to the same conclusions: A strong and significant effect of land use and a higher variability of the community structure was observed for samples from scrubland (likely due to the absence of mixing). Furthermore, compared to the effect of land use, the effect of the site was less pronounced.

Tillage is probably one of the major forces driving shifts in the soil microbial community structure. In addition, irrigation, fertilization, and/or application of agrochemicals have been shown to affect the bacterial community structure [14], [39]–[42]. The minor differences found between the arable soils from site 1 and site 2 may be linked to differences in their previous cropping history and agricultural management, which were most likely not identical. In general, the conversion of scrubland into agricultural land is associated with the replacement of a diverse, indigenous, highly adapted vegetation of the semi-arid desert with a few crop plants. Plant effects on soil microbial communities have been frequently observed in the rhizosphere, which refers to the soil directly influenced by root exudates [30], [43]. The influence of plant species varied between different studies, in some of them it was regarded to be the major factor shaping the microbial community structure [44]–[49], while in others only a minor influence of plant species and vegetation composition on the soil bacterial community composition was observed [50]. However, only a few studies suggested long-term effects of plants on microbial communities in bulk soils [16]. Long-term agricultural use impacts soil physicochemical characteristics [3], [16], [17], and thus probably alters the composition and properties of biogeochemical interfaces in soils [51].

The present study showed that agricultural use impacted several bacterial phyla in the soils of the sites studied, e.g., *Acidobacteria, Chloroflexi, Spirochaetes, Verrucomicrobia, Gemmatimonadetes, Deltaproteobacteria* (*Proteobacteria*), *Acidimicrobiales* (*Actinobacteria*) and *Rubrobacterales* (*Actinobacteria*). *Acidobacteria* have been found to be dominant in several soils [52] though often they are difficult to cultivate. In the study by Bisette et al. [16], the proportion of *Acidobacteria* was found increased in grassland soils compared to soils under agricultural use at one site in Australia. Compared with agricultural soils, their relative abundance was also reported to be higher in forest, desert or prairie soils [52], [53]. The proportion of *Acidobacteria* was reported to be significantly lower in nutrient-rich rhizosphere than in bulk soil, confirming their oligotrophic lifestyle [54]. Under dry conditions, the net primary productivity of plants is controlled by water [55], [56]. Additional water supply in the alfalfa fields enhanced plant growth and probably elevated the extent of plant exudates into soil over a time period of about 50 years. Recently the relative abundance of *Acidobacteria* was found to be negatively correlated with the level of nitrogen input (fertilizer) [57], which in general increases the net productivity of vegetation. *Chloroflexi* also was reported to prevail in nutrient poor soils [52], [58] and other oligotrophic ecosystems such as soils from high-elevation regions where vegetation is patchy [59], alpine tundra soil [60] or hyperarid polar desert soil [61]. A few bacteria belonging to *Chloroflexi* which could be retrieved from soil had very slow rates of growth and mini-colony formation [62], [63] which are typical characteristics of oligotrophic organisms. In accordance with the present study, Fierer et al. [57] showed that the relative abundance of *Chloroflexi* was also lower in the plots with high levels of nitrogen input.

In the present study, the physicochemical analysis of the soils done for four independent replicates per site and treatment revealed that the concentration of sulphate (probably due to fertilization) was higher in alfalfa soils. Sulphate can be used as terminal electron acceptor by some anaerobic bacteria. However, a high proportion of OTUs belonging to *Desul-*

fobacterales, Desulfovibrionales, Syntrophobacterales or *Clostridia* had significantly lower abundances in the alfalfa soils compared to scrubland soils. By and large, bacteria belonging to these taxonomic groups are strict anaerobes [64] and play an important role in anaerobic carbon cycling in wet terrestrial ecosystems such as rice fields and wet lands [65,66). Furthermore, tilling soils also affects soil aeration [67] and can elevate the activity of aerobic microbes [14]. Notably, in the present study, a significant loss of organic matter also was observed in the agricultural soils and as soil organic matter is an important binding agent of soil particles into aggregates [68], presumably lowering gas diffusion, this may have contributed to lower sulphate-reducing populations in these desert agricultural soils.

High proportions of OTUs belonging to the *Bacteroidetes* or the *Alpha-, Beta-* and *Gammaproteobacteria* as well as several families of *Actinobacteria* were more abundant in alfalfa soils. The higher abundance of *Pseudomonas* (50% of the OTUs with significantly higher signal intensities) in the agricultural soils was also confirmed by the detection of *Pseudomonas*-specific gacA genes [46]. Amplicons of the gacA gene fragment were obtained only for TC DNA from alfalfa soils, not from scrubland soil (data not shown). *Betaproteobacteria* and *Bacteroides* were considered to contain copiotrophic taxa as their abundance has been positively correlated with carbon mineralization rate and carbon availability [69]. Several studies reported enrichments of *Alphaproteobacteria* [69], *Gammaproteobacteria* [46,70) or *Actinobacteria* [29,43) in the rhizosphere where carbon availability is increased due to root exudates. The number of OTUs detected by PhyloChip analysis was significantly lower in the soil from scrubland than alfalfa field. In general, this finding is in agreement with other studies, in which agriculture or low plant diversity (often a direct result of converting natural land into agricultural use) did not necessarily lead to a reduction of the bacterial diversity detected [48], [71]–[73]. However, the effects of agricultural practices on soil bacterial richness remain to be explored more fully, perhaps at a finer resolution than was obtainable in this study. Firstly, total bacterial species richness in soils is difficult to assess as many populations occur at low abundance [74], [75], [76]. Secondly, a high proportion of OTUs belonging to *Acidobacteria, Chloroflexi, Spirochaetes, Verrucomicrobia, Gemmatimonadetes* were more abundant in the scrubland soils, suggesting that these taxa were better adapted to those soils. Compared to *Proteobacteria, Actinobacteria* and *Firmicutes*, many fewer OTUs belonging to these taxa were available during probe design. Therefore, bacterial richness of these phyla in the scrubland soil possibly was underestimated.

In summary, although investigated only at two sites that were assumed to be representative for this type of ecosystem, we could show that the use of scrublands for agriculture caused profound changes in the soil bacterial community structure and physicochemical characteristics. Soil parameters that differed between land uses were highly correlated with the community composition of taxa responding to land use. Several, most likely oligotrophic or anaerobic taxa were negatively affected by the change, in contrast, populations with a potentially copiotrophic lifestyle profited and were enhanced in the agricultural soils.

REFERENCES

1. Munson SM, Belnap J, Okin GS (2011) Responses of wind erosion to climate-induced vegetation changes on the Colorado Plateau. Proc Natl Acad Sci U S A 108: 3854–3859. doi: 10.1073/pnas.1014947108
2. Saygin SD, Basaran M, Ozcan AU, Dolarslan M, Timur OB, et al. (2011) Land degradation assessment by geo-spatially modeling different soil erodibility equations in a semi-arid catchment. Environ Monit Assess 180: 201–215. doi: 10.1007/s10661-010-1782-z
3. Zhao WZ, Xiao HL, Liu ZM, Li J (2005) Soil degradation and restoration as affected by land use change in the semiarid Bashang area, northern China. Catena 59: 173–186. doi: 10.1016/j.catena.2004.06.004
4. Mahamane L, Mahamane S (2005) Biodiversity of ligneous species in semi-arid to arid zones of southwestern Niger according to anthropogenic and natural factors. Agr Ecosyst Environ 105: 267–271. doi: 10.1016/j.agee.2004.03.004
5. Underwood EC, Viers JH, Klausmeyer KR, Cox RL, Shaw MR (2009) Threats and biodiversity in the mediterranean biome. Divers Distrib 15: 188–197. doi: 10.1111/j.1472-4642.2008.00518.x
6. Foley JA, DeFries R, Asner GP, Barford C, Bonan G, et al. (2005) Global consequences of land use. Science 309: 570–574. doi: 10.1126/science.1111772
7. Snyman HA (1998) Dynamics and sustainable utilization of rangeland ecosystems in arid and semi-arid climates of southern Africa. J Arid Environ 39: 645–666. doi: 10.1006/jare.1998.0387

8. Maestre FT, Cortina J, Vallejo R (2006) Are ecosystem composition, structure, and functional status related to restoration success? A test from semiarid Mediterranean steppes. Restor Ecol 14: 258–266. doi: 10.1111/j.1526-100x.2006.00128.x
9. Choudhary DK, Sharma KP, Gaur RK (2011) Biotechnological perspectives of microbes in agro-ecosystems. Biotechnol Lett 33: 1905–1910. doi: 10.1007/s10529-011-0662-0
10. Miransari M (2011) Soil microbes and plant fertilization. Appl Microbiol Biotechnol 92: 875–885. doi: 10.1007/s00253-011-3521-y
11. Uroz S, Calvaruso C, Turpault MP, Frey-Klett P (2009) Mineral weathering by bacteria: ecology, actors and mechanisms. Trends Microbiol 17: 378–387. doi: 10.1016/j.tim.2009.05.004
12. Bell T, Newman JA, Silverman BW, Turner SL, Lilley AK (2005) The contribution of species richness and composition to bacterial services. Nature 436 1157–1160. doi: 10.1038/nature03891
13. Giovannoni S (2004) Evolutionary biology: oceans of bacteria. Nature 430: 515–516. doi: 10.1038/430515a
14. Calderon FJ, Jackson LE, Scow KM, Rolston DE (2001) Short-term dynamics of nitrogen, microbial activity, and phospholipid fatty acids after tillage. Soil Sci Soc Am J 65: 118–126. doi: 10.2136/sssaj2001.651118x
15. Zak DR, Holmes WE, White DC, Peacock AD, Tilman D (2003) Plant diversity, soil microbial communities, and ecosystem function: Are there any links? Ecology 84: 2042–2050. doi: 10.1890/02-0433
16. Bissett A, Richardson AE, Baker G, Thrall PH (2011) Long-term land use effects on soil microbial community structure and function. Appl Soil Ecol 51: 66–78. doi: 10.1016/j.apsoil.2011.08.010
17. Peixoto RS, Chaer GM, Franco N, Reis FB, Mendes IC, et al. (2010) A decade of land use contributes to changes in the chemistry, biochemistry and bacterial community structures of soils in the Cerrado. Antonie Van Leeuwenhoek 98: 403–413. doi: 10.1007/s10482-010-9454-0
18. Salinas-Zavala CA, Lluch-Cota SE, Fogel I (2006) Historic development of winter-wheat yields in five irrigation districts in the Sonora Desert, Mexico. Interciencia 31: 254–261.
19. Coria R, Leon de la Luz JS (1992) Flora iconografica de Baja California Sur. Centro de Investigaciones Biologicas de Baja California Sur A.C.
20. Armenta-Quintana JA, Ramirez-Orduna R, Ramirez-Lozano RG (2011) Forage utilization and diet selection by grazing goats on a sarcocaulescent scrubland in Northwest Mexico. Rev Chapingo Ser Cie 17: 163–171.
21. Ramirez-Orduna R, Ramirez RG, Romero-Vadillo E, Gonzalez-Rodriguez H, Armenta-Quintana JA, et al. (2008) Diet and nutrition of range goats on a sarcocaulescent shrubland from Baja California Sur, Mexico. Small Ruminant Res 76: 166–176. doi: 10.1016/j.smallrumres.2007.12.020
22. Gee GW, Or D (2002) Particle size analysis. In: Dane JH, Topp GC (eds). Methods of Soil Analysis. Part 4-Physical Methods, Soil Sci. Soc., America, Madison, 255–293.

23. McLean EO (1982) Soil pH and lime requirement. In: Page, AL (ed). Methods of Soil Analysis, Part 2, American Soc. Agronomy, Madison, 199–224.
24. Nelson DE, Sommers LE (1996) Total organic carbon and organic matter. In: Page, AL (ed). Methods of Soil Analysis. Part 2, American Soc. Agronomy, Madison, 961–1010.
25. Strickland JDH, Parsons TR (1972) A practical handbook of seawater analysis. Fisheries Res. Board Canada, Ottawa, Canada.
26. Jackson ML (1958) Soil Chemical Analysis. Prentice Hall, Englewood, N.J.
27. Olsen SR, Dean LA (1965) Phosphorus. Chemical and microbiological properties. In: Black, CA (ed). Methods of Soil Analysis, Part 2. Am. Soc. Agron., Madison, WI, USA, 1035–1048.
28. Heuer H, Krsek M, Baker P, Smalla K, Wellington EM (1997) Analysis of actinomycete communities by specific amplification of genes encoding 16S rRNA and gel-electrophoretic separation in denaturing gradients. Appl Environ Microbiol 63: 3233–3241.
29. Gomes NCM, Heuer H, Schönfeld J, Costa R, Mendonça-Hagler L, et al. (2001) Bacterial diversity of the rhizosphere of maize (Zea mays) grown in tropical soil studied by temperature gradient gel electrophoresis. Plant Soil 232: 167–180. doi: 10.1007/978-94-010-0566-1_17
30. Weinert N, Meincke R, Gottwald C, Heuer H, Gomes NC, et al. (2009) Rhizosphere communities of genetically modified zeaxanthin-accumulating potato plants and their parent cultivar differ less than those of different potato cultivars. Appl Environ Microbiol 75: 3859–3865. doi: 10.1128/aem.00414-09
31. Gomes NCM, Kosheleva IA, Abraham WR, Smalla K (2005) Effects of the inoculant strain Pseudomonas putida KT2442 (pNF142) and of naphthalene contamination on the soil bacterial community. FEMS Microbiol Ecol 54: 21–33. doi: 10.1016/j.femsec.2005.02.005
32. Heuer H, Wieland J, Schönwälder A, Gomes NCM, Smalla K (2001) Bacterial community profiling using DGGE or TGGE analysis. In: Rouchelle, IP (ed). Environmental molecular microbiology: protocols and applications. Horizon Scientific Press: Wymondham, United Kingdom. 177–190.
33. Kropf S, Heuer H, Grüning M, Smalla K (2004) Significance test for comparing complex microbial community fingerprints using pairwise similarity measures. J Microbiol Methods 57: 187–195. doi: 10.1016/j.mimet.2004.01.002
34. Weinert N, Piceno Y, Ding GC, Meincke R, Heuer H, et al. (2011) PhyloChip hybridization uncovered an enormous bacterial diversity in the rhizosphere of different potato cultivars: many common and few cultivar-dependent taxa. FEMS Microbiol Ecol 75: 497–506. doi: 10.1111/j.1574-6941.2010.01025.x
35. DeSantis TZ, Brodie EL, Moberg JP, Zubieta IX, Piceno YM, et al. (2007) High-density universal 16S rRNA microarray analysis reveals broader diversity than typical clone library when sampling the environment. Microb Ecol 53: 371–383. doi: 10.1007/s00248-006-9134-9
36. Brodie EL, DeSantis TZ, Parker JP, Zubietta IX, Piceno YM, et al. (2007) Urban aerosols harbor diverse and dynamic bacterial populations. Proc Natl Acad Sci USA 104: 299–304. doi: 10.1073/pnas.0608255104
37. Ivanov II, Atarashi K, Manel N, Brodie EL, Shima T, et al. (2009) Induction of intestinal Th17 cells by segmented filamentous bacteria. Cell 139: 485–498. doi: 10.1016/j.cell.2009.09.033

38. Ding GC, Smalla K, Heuer H, Kropf S (2012) A new proposal for a principal component-based test for high-dimensional data applied to the analysis of PhyloChip data. Biom J 54: 94–107. doi: 10.1002/bimj.201000164
39. Entry JA, Fuhrmann JJ, Sojka RE, Shewmaker GE (2004) Influence of irrigated agriculture on soil carbon and microbial community structure. Environ Manage 33: S363–S373. doi: 10.1007/s00267-003-9145-y
40. Islam MR, Trivedi P, Palaniappan P, Reddy MS, Sa T (2009) Evaluating the effect of fertilizer application on soil microbial community structure in rice based cropping system using fatty acid methyl esters (FAME) analysis. World J Microb Biot 25: 1115–1117. doi: 10.1007/s11274-009-9959-8
41. Lo CC (2010) Effect of pesticides on soil microbial community. J Environ Sci Health B 45: 348–359. doi: 10.1080/03601231003799804
42. Ramirez KS, Lauber CL, Knight R, Bradford MA, Fierer N (2010) Consistent effects of nitrogen fertilization on soil bacterial communities in contrasting systems. Ecology 91: 3463–3470. doi: 10.1890/10-0426.1
43. Smalla K, Wieland G, Buchner A, Zock A, Parzy J, et al. (2001) Bulk and rhizosphere soil bacterial communities studied by denaturing gradient gel electrophoresis: Plant-dependent enrichment and seasonal shifts revealed. Appl Environ Microb 67: 4742–4751. doi: 10.1128/aem.67.10.4742-4751.2001
44. Costa R, Götz M, Mrotzek N, Lottmann J, Berg G, et al. (2006) Effects of site and plant species on rhizosphere community structure as revealed by molecular analysis of microbial guilds. FEMS Microbiol Ecol 56: 236–249. doi: 10.1111/j.1574-6941.2005.00026.x
45. Costa R, Salles JF, Berg G, Smalla K (2006) Cultivation-independent analysis of Pseudomonas species in soil and in the rhizosphere of field-grown Verticillium dahliae host plants. Environ Microbiol 8: 2136–2149. doi: 10.1111/j.1462-2920.2006.01096.x
46. Costa R, Gomes NC, Krögerrecklenfort E, Opelt K, Berg G, et al. (2007) Pseudomonas community structure and antagonistic potential in the rhizosphere: insights gained by combining phylogenetic and functional gene-based analyses. Environ Microbiol 9: 2260–2273. doi: 10.1111/j.1462-2920.2007.01340.x
47. DeAngelis KM, Brodie EL, DeSantis TZ, Andersen GL, Lindow SE, et al. (2009) Selective progressive response of soil microbial community to wild oat roots. ISME J 3: 168–178. doi: 10.1038/ismej.2008.103
48. Kowalchuk GA, Buma DS, de Boer W, Klinkhamer PG, van Veen JA (2002) Effects of above-ground plant species composition and diversity on the diversity of soil-borne microorganisms. Antonie Van Leeuwenhoek 81: 509–520. doi: 10.1023/a:1020565523615
49. Mendes R, Kruijt M, de Bruijn I, Dekkers E, van der Voort M, et al. (2011) Deciphering the rhizosphere microbiome for disease-suppressive bacteria. Science 332: 1097–1100. doi: 10.1126/science.1203980
50. Kielak A, Pijl AS, van Veen JA, Kowalchuk GA (2008) Differences in vegetation composition and plant species identity lead to only minor changes in soil-borne microbial communities in a former arable field. FEMS Microbiol Ecol 63: 372–382. doi: 10.1111/j.1574-6941.2007.00428.x
51. Totsche KU, Rennert T, Gerzabek MH, Kögel-Knabner I, Smalla K, et al. (2010) Biogeochemical interfaces in soil: The interdisciplinary challenge for soil science. J Plant Nutr Soil Sci 173: 88–99. doi: 10.1002/jpln.200900105

52. Janssen PH (2006) Identifying the dominant soil bacterial taxa in libraries of 16S rRNA and 16S rRNA genes. Appl Environ Microbiol 72: 1719–1728. doi: 10.1128/aem.72.3.1719-1728.2006
53. Fierer N, Jackson JA, Vilgalys R, Jackson RB (2005) Assessment of soil microbial community structure by use of taxon-specific quantitative PCR assays. Appl Environ Microbiol 71: 4117–4120. doi: 10.1128/aem.71.7.4117-4120.2005
54. Kielak A, Pijl AS, van Veen JA, Kowalchuk GA (2009) Phylogenetic diversity of Acidobacteria in a former agricultural soil. ISME J 3: 378–382. doi: 10.1038/ismej.2008.113
55. Gao YZ, Chen Q, Lin S, Giese M, Brueck H (2011) Resource manipulation effects on net primary production, biomass allocation and rain-use efficiency of two semiarid grassland sites in Inner Mongolia, China. Oecologia 165: 855–864. doi: 10.1007/s00442-010-1890-z
56. Webb W, Szarek S, Lauenroth W, Kinerson R, Smith M (1978) Primary productivity and water-use in native forest, grassland, and desert ecosystems. Ecology 59: 1239–1247. doi: 10.2307/1938237
57. Fierer N, Lauber CL, Ramirez KS, Zaneveld J, Bradford MA, et al. (2012) Comparative metagenomic, phylogenetic and physiological analyses of soil microbial communities across nitrogen gradients. ISME J 6: 1007–1017. doi: 10.1038/ismej.2011.159
58. Will C, Thurmer A, Wollherr A, Nacke H, Herold N, et al. (2010) Horizon-specific bacterial community composition of German grassland soils, as revealed by pyrosequencing-based analysis of 16S rRNA genes. Appl Environ Microbiol 76: 6751–6759. doi: 10.1128/aem.01063-10
59. Freeman KR, Pescador MY, Reed SC, Costello EK, Robeson MS, et al. (2009) Soil CO2 flux and photoautotrophic community composition in high-elevation, 'barren' soil. Environ Microbiol 11: 674–686. doi: 10.1111/j.1462-2920.2008.01844.x
60. Costello EK, Schmidt SK (2006) Microbial diversity in alpine tundra wet meadow soil: novel Chloroflexi from a cold, water-saturated environment. Environ Microbiol 8: 1471–1486. doi: 10.1111/j.1462-2920.2006.01041.x
61. Pointing SB, Chan YK, Lacap DC, Lau MCY, Jurgens JA, et al. (2009) Highly specialized microbial diversity in hyper-arid polar desert. Proc Natl Acad Sci USA 106: 19964–19969. doi: 10.1073/pnas.0908274106
62. Davis KE, Joseph SJ, Janssen PH (2005) Effects of growth medium, inoculum size, and incubation time on culturability and isolation of soil bacteria. Appl Environ Microbiol 71: 826–834. doi: 10.1128/aem.71.2.826-834.2005
63. Davis KER, Sangwan P, Janssen PH (2011) Acidobacteria, Rubrobacteridae and Chloroflexi are abundant among very slow-growing and mini-colony-forming soil bacteria. Environ Microbiol 13: 798–805. doi: 10.1111/j.1462-2920.2010.02384.x
64. Muyzer G, Stams AJ (2008) The ecology and biotechnology of sulphate-reducing bacteria. Nat Rev Microbiol 6: 441–454. doi: 10.1038/nrmicro1892
65. He JZ, Liu XZ, Zheng Y, Shen JP, Zhang LM (2010) Dynamics of sulfate reduction and sulfate-reducing prokaryotes in anaerobic paddy soil amended with rice straw. Biol Fert Soils 46: 283–291. doi: 10.1007/s00374-009-0426-3

66. Pester M, Knorr KH, Friedrich MW, Wagner M, Loy A (2012) Sulfate-reducing microorganisms in wetlands - fameless actors in carbon cycling and climate change. Front Microbiol 3: 72. doi: 10.3389/fmicb.2012.00072
67. Kladivko EJ (2001) Tillage systems and soil ecology. Soil Till Res 61: 61–76. doi: 10.1016/s0167-1987(01)00179-9
68. Bronick CJ, Lal R (2005) Soil structure and management: a review. Geoderma 124: 3–22. doi: 10.1016/j.geoderma.2004.03.005
69. Fierer N, Bradford MA, Jackson RB (2007) Toward an ecological classification of soil bacteria. Ecology 88: 1354–1364. doi: 10.1890/05-1839
70. Rossmann B, Muller H, Smalla K, Mpiira S, Tumuhairwe JB, et al. (2012) Banana-associated microbial communities in Uganda are highly diverse but dominated by Enterobacteriaceae. Appl Environ Microbiol 78: 4933–4941. doi: 10.1128/aem.00772-12
71. Gruter D, Schmid B, Brandl H (2006) Influence of plant diversity and elevated atmospheric carbon dioxide levels on belowground bacterial diversity. BMC Microbiol 6: 68.
72. Koberl M, Muller H, Ramadan EM, Berg G (2011) Desert farming benefits from microbial potential in arid soils and promotes diversity and plant health. PLoS One 6: e24452. doi: 10.1371/journal.pone.0024452
73. Zul D, Denzel S, Kotz A, Overmann J (2007) Effects of plant biomass, plant diversity, and water content on bacterial communities in soil lysimeters: implications for the determinants of bacterial diversity. Appl Environ Microbiol 73: 6916–6929. doi: 10.1128/aem.01533-07
74. Curtis TP, Sloan WT, Scannell JW (2002) Estimating prokaryotic diversity and its limits. Proc Natl Acad Sci U S A 99: 10494–10499. doi: 10.1073/pnas.142680199
75. Gans J, Wolinsky M, Dunbar J (2005) Computational improvements reveal great bacterial diversity and high metal toxicity in soil. Science 309: 1387–1390. doi: 10.1126/science.1112665
76. Hong SH, Bunge J, Jeon SO, Epstein SS (2006) Predicting microbial species richness. Proc Natl Acad Sci U S A 103: 117–122. doi: 10.1073/pnas.0507245102

There are several supplemental files that are not available in this version of the article. To view this additional information, please use the citation information cited on the first page of this chapter.

PART III

ENERGY AND GREENHOUSE GASES

CHAPTER 6

A GREENHOUSE GAS AND SOIL CARBON MODEL FOR ESTIMATING THE CARBON FOOTPRINT OF LIVESTOCK PRODUCTION IN CANADA

XAVIER P.C. VERGÉ, JAMES A. DYER, DEVON E. WORTH, WARD N. SMITH, RAYMOND L. DESJARDINS, AND BRIAN G. MCCONKEY

6.1 INTRODUCTION

The growing global demand for food will compete with efforts to mitigate Greenhouse Gas (GHG) emissions and adapt to climate change [1,2]. With the expansion of high protein diet among many emerging economies, large land areas will be required for livestock feed production [3,4]. This land use will compete with crops for direct human consumption or biofuel feedstock [5,6]. There is also concern about the large amounts of enteric methane emitted by ruminant livestock [7], as well as the emissions of nitrous oxide and carbon dioxide from all types of livestock operations [8]. Important questions arise about which types of livestock satisfy the demand for protein most efficiently, make the best use of the land resource base and have the lowest carbon footprint. Therefore, to help the live-

This chapter was originally published under the Creative Commons Attribution License. Vergé XPC, Dyer JA, Worth DE, Smith WN, Desjardins RL and McConkey BG. A Greenhouse Gas and Soil Carbon Model for Estimating the Carbon Footprint of Livestock Production in Canada. Animals ***2012*** *,2 (2012); 437-454. doi:10.3390/ani2030437.*

stock industries cope with these pressures, an objective set of algorithms that can compare how various livestock types impact the environment and meet growing food demands will be needed.

Sequestering atmospheric carbon dioxide (CO_2) as soil carbon is a potential strategy for reducing GHG emissions [9]. About one third of the soil carbon stock was lost when virgin soils were first broken with the plow in Canada and this stock declined further under continued mechanized cultivation [10]. Better farming practices over the last twenty years, however, have reversed this trend. Canada's agricultural soils, which were estimated to have been a small CO_2 source in 1991, were a small sink by 2001 [11]. But there are tradeoff effects with regard to the extent to which all of these practices are applied and which crops are grown. For example, increases in nitrous oxide (N_2O) and methane (CH_4) emissions from livestock and crop productions have been offset by carbon sequestration in soils [11]. Forage crops enhance soil organic matter, but feeding those crops to livestock increase CH_4 emissions [12].

The main objective of this paper was to present a dynamic, quantitative model for estimating GHG emissions and determining the carbon footprint of Canadian livestock industries. To determine the long term impact of changes in livestock populations on the carbon footprint of Canadian farms, this model integrates the changes in soil carbon associated with shifts among livestock populations with their annual GHG emission budgets. The second objective was to demonstrate the linkages among GHG emission calculations that are livestock type specific and their relationship with soil carbon. To achieve the second objective four scenarios for livestock industry interactions (described below) were assumed.

6.2 EXPERIMENTAL SECTION

6.2.1 DEVELOPMENT OF A LIVESTOCK GHG MODEL

6.2.1.1 BACKGROUND

Commodity assessments have been previously completed for the Canadian dairy, beef pork and poultry industries [13–16]. However, these assessments did not consider inter-commodity interactions. Satisfying livestock diet requirements can lead to competition for feed grain, especially under intensive animal production. Since livestock industries must share arable land with food crops, oilseeds [6] and biofuel feedstock crops [5,17], livestock industries can no longer be treated as separate closed systems. The Unified Livestock Industry and Crop Emissions Estimation System (ULICEES) was created by assembling the four sets of livestock GHG computations in one model. ULICEES also takes the changes in the soil carbon stock into account. Although ULICEES is applicable to any agricultural census year, in this analysis it was applied to 2001, the most recent year for which the livestock diet survey data was available [18].

Figure 1 presents the generalized computational flow of the set of calculations for quantifying the GHG emission budgets for Canadian livestock production. Only the functions common among all four livestock industries are illustrated. The yields, areas, and fertilizer application rates for each crop and population for each livestock type are shown as computational inputs. The resulting GHG emission categories and totals are shown at the bottom of the chart. Although ULICEES addresses the question of how to measure the carbon footprint of all food of animal origin in Canada [8], this chart does not show interactions among commodities and it only considers changes in soil CO_2 emissions under a land use change. All of the GHGs associated with animal housing [19] were also taken into account in ULICEES.

6.2.1.2 CROP COMPLEX

The scope of assessment adopted by Vergé et al. [14] included the associated complex of crops that supported each commodity production system. The carbon footprint of animal based production cannot be effectively quantified without first determining the GHG emissions from growing the feed grains and the forage they consume. That land base, defined as the livestock crop complex (LCC), was the result of integrating the diets of all the age-gender categories in each livestock type over the whole population. Each crop component of each diet was divided by the average yield of that crop to estimate the land required to grow the crop.

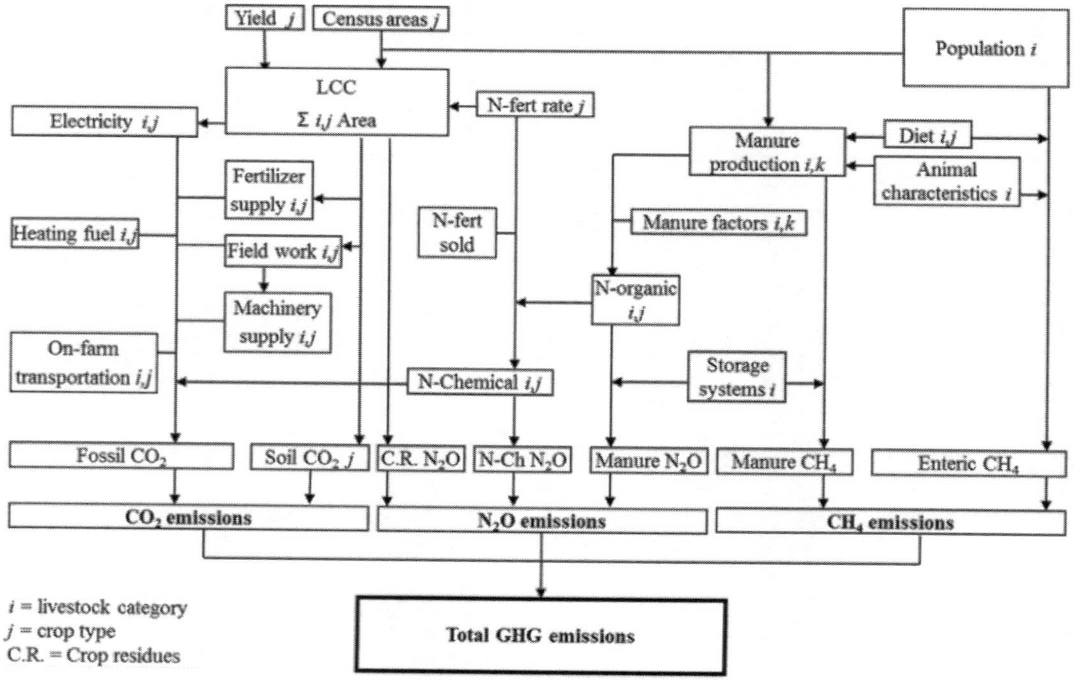

FIGURE 1: Chart of the generalized computational flow of the Greenhouse Gas (GHG) emission budgets of the four major Canadian livestock industries.

Specific crop complexes, DCC, BCC, PCC and ACC, were defined for the dairy, beef, pork and poultry (avian) industries, respectively, in Canada [4,13–16]. The LCC was a critical step in integrating livestock commodities in ULICEES. Livestock ration records obtained from Statistics Canada [18] were combined with animal population data from agriculture census records [20,21] to calculate the LCC of each livestock industry. By defining the area needed to feed all on-farm animals, the LCC concept sets limits on the livestock production system. It recognizes that the GHG emissions attributed to livestock do not stop at the feedlot or the animals. Even if farmers buy all their livestock feed, the GHGs emitted from the land on which those crops were grown were still attributed to those livestock.

6.2.1.3 THE GHG EMISSIONS AND SOURCES

Like the four previous commodity specific GHG emission assessments, ULICEES accounted for CH_4, N_2O and fossil CO_2. Since these emission calculations were described in detail in the previous assessments [4,14–16], only a general overview of these calculations is described here. Enteric methane emissions from ruminants were based on the Intergovernmental Panel on Climate Change (IPCC) Tier 2 methodology [22], adapted for Canadian conditions by Vergé et al. [23]. These emission estimates also accounted for a small but measurable amount of enteric methane from hogs [24]. The default IPCC tier 1 emission factors were used and corrected based on the animal weight as presented by Vergé et al. [15]. Methane emissions from manure were calculated using the IPCC Tier 2 methodology [22] for all animal types. Methane emission factors for each age gender category were then multiplied by their respective populations.

Based on the IPCC tier 1 methodology the total amount of nitrogen applied in each LCC was used to define the N_2O emissions from each livestock production system [22,25]. The N_2O emission factors adapted to Canadian conditions came from Rochette et al. [26]. Different computation pathways were required depending on whether the nitrogen source was organic or chemical. The annual amount of organic nitrogen applied was based on the quantity of manure produced by each livestock population and the nitrogen content of each type of manure. The amount of chemical nitrogen fertilizer applied was obtained by subtracting the organic nitrogen from the total required nitrogen. Crop specific nitrogen fertilizer application recommendations [27] were integrated over the respective LCC crop areas to derive the total nitrogen in the LCC. The amount of chemical nitrogen was adjusted to the amount of fertilizer actually sold in each region [28]. This adjustment factor was calculated by comparing the total recommended amount of nitrogen calculated and assumed to be applied within each province to the total nitrogen fertilizer purchased in the same province [4,14–16].

Although smaller in magnitude than agricultural methane and nitrous oxide emissions, farm energy is an essential part of the sector's GHG emissions budget [29,30]. A combination of farm statistics and agricultural engineering coefficients were used to estimate the fossil CO_2 emissions from the fossil fuel for farm fieldwork [31,32]. The energy and fossil CO_2 emissions associated with on farm transport, farm use of electricity and heating fuel and the indirect fossil energy to manufacture and transport farm machinery and chemical fertilizer to the farm were also included [30,33,34].

6.2.1.4 THE SOIL CARBON STOCK

In previous commodity specific applications of the LCC methodology, CO_2 emissions from soil carbon were not considered because each livestock system was treated in isolation with little change in overall land use or management. ULICEES treats soil carbon as an exhaustible storage term, including changes in soil organic carbon (SOC) between different land-use management systems in a similar manner as IPCC GHG accounting methodology [22], whereby a land-use/management system that is in equilibrium is converted to a new land-use/management system and is assumed to reach a new equilibrium within a time frame. Some forage fields that are in rotation with annual crops may not be at equilibrium. For these cases, additional atmospheric CO_2 which would have been sequestered if those forage areas in rotation had been given time to come to equilibrium, along with the sequestered soil carbon. Including that lost sequestration potential would, therefore, result in the same total loss of sequestered CO2 as lost from the soil under continuous perennial forage.

ULICEES assumes a new SOC equilibrium after 40 years. We assumed that after 40 years SOC stabilizes to a new soil organic carbon level where it no longer makes an appreciable contribution to the annual GHG emissions budget. As well, 40 years from the present, 2050, is approximately the time when GHG levels in the atmosphere were projected at the Nairobi Climate Change Summit (COP 12) to double compared to pre-industrial atmospheric levels [35]. For each livestock production system, yearly GHG emissions of CH_4, N_2O and fossil CO_2 continue indefinitely as long as that production system operates. Carbon flow is, however, not permanent or linear from a long term carbon balance perspective [11].

6.2.2 THE PAYBACK PERIOD

Based on the difference in GHG emissions from beef and pork, a period can be calculated that is required to accumulate a multi-year quantity of GHG emissions that equal the difference in carbon sequestration between the two land uses. The time required to compensate for an amount of soil carbon that is lost as a result of a land change is defined as the payback period. This payback period relates the loss of soil carbon to the annual GHG emissions. A precedent has been set for the payback period approach in life cycle assessments of biofuels [36–39]. For example, if a tropical forest is cut down to grow palm oil for biodiesel feedstock, it would take many decades before the annual offset of fossil CO_2 emissions from the biodiesel equals that loss in tropical soil carbon. The beef to pork conversion can be treated in the same way. The change in crop areas for the beef to pork redistribution would come from the forage area that supported the displaced population of beef which would then be converted to feed grain for additional hogs.

Since the complete decay curve for soil carbon between two steady states is exponential, its slope approaches zero asymptotically near the equilibrium. Hence, the decay period must be defined in terms of a tolerable amount of residual soil carbon. In this analysis, the decay period was set at 40 years which accounts for about 60% of the carbon stock [40]. The remaining 40% would be lost over the next 60 years at an average annual rate that is less than half of the average soil carbon decay rate over the first 40 years. The integrated annual GHG emissions over the payback period can be compared to the change in soil carbon stock over that period. A payback period that is appreciably shorter than the 40 year decay period represents a net gain in GHG mitigation potential.

6.2.3 THE BEEF TO PORK REDISTRIBUTION

To demonstrate the inter-commodity interactions, a potential expansion of the pork industry was assumed which would displace some of the beef industry in Canada. Whereas beef cattle in Canada are raised mainly on roughages supplemented by grain, hogs are completely dependent on annual crops. Therefore, additional feed grain area would be required for the expanded hog population. This livestock redistribution illustrates the tradeoff between the reduction of enteric methane emissions and the soil carbon loss during the replacement of perennial forage cover with annual crops. Related changes in the LCC will include more fossil fuel use for farm field operations and an increase in N_2O emissions due to higher nitrogen fertilizer requirements.

The quantitative basis of the beef to pork redistribution test was to avoid any loss of protein supply. Hence, the increase in pork production must supply the same amount of protein as was lost from beef production. The two quantities of protein were calculated using the protein to live weight conversion factors from Dyer et al. [41]. The beef and pork systems have different intensities based on protein production [41], These differences in GHG emission intensities can also be seen in beef and pork comparisons based on live weight production [4,42–44]. However, it was not the objective of this paper to compare productivities of these industries, but to ensure that the loss in food production from the land use change would be minimized.

Because the BCC can include land that is typically only suitable for growing perennial forages [45], only a portion of the BCC can be reallocated to grow annual crops in the PCC to feed hogs. As recognized by Basarab et al. [44], the chance of some of that redistributed land being of too low quality to grow the additional feed grains for hogs had to be minimized. By transferring only 10% of the beef based protein production potential to pork production, only the best portion of the BCC was allowed to be involved in the redistribution of land in both eastern and western Canada. The increase in pork production as a percentage of the total annual supply of protein from pork must be larger than 10% where the supply of protein from pork is less than the total amount of protein from beef. It must be less than 10% where the pork protein supply exceeds the supply of protein from beef. Whereas the deflation factor for beef was 90% for both eastern and western Canada, the corresponding inflation factors for pork were different between the two Canadian regions (Table 1).

TABLE 1: Weights of protein before and after redistribution from beef to pork production and the beef deflation and pork inflation factors.

	Initial	Reallocated	Remaining	After/before factors (%)
		kt, protein		
		Beef		for deflation
East	36.9	3.7	33.2	90
West	218.8	21.9	197.0	90
		Pork		for inflation
East	157.7	3.7	161.4	102
West	123.5	21.9	145.4	118

The beef to pork redistribution involves three crop area changes: area going into feed grains for pork (Ap); the forage area that was supporting displaced beef (Af); and the feed grain area that was supporting displaced beef (Ag). Some of the area required to grow feed grain for the hogs had to be taken from land that had been growing forage. The conversion from forage to annual grains determined the area (cA) in which the changes in soil carbon storage caused solely by the expansion of pork production take place. The initial land displacement was computed as:

$$\Delta cA = Ap - Ag \quad (1)$$

Not all of the land in forage was needed to expand the feed grain crop to support more hogs. Being a more extensive system, there will be land left over from the beef industry as the required perennial forage is reduced. The GHG emissions of the residual forage area (rA) could undergo a further change because it is no longer needed by the displaced beef cattle. The residual area from the initial land displacement was computed as:

$$\Delta rA = Af - (Ap - Ag) \quad (2)$$

6.2.4 RESIDUAL LAND REDISTRIBUTION SCENARIOS

It is difficult to predict with any certainty the most likely land use for ΔrA from the beef to pork redistribution. Because of this uncertainty, four scenarios for the residual forage area were examined. These scenarios result in a range of impacts on the total carbon footprint of the redistribution from the perspective of the land use and crop choices for ΔrA.

- Scenario 1 assumes that ΔrA will remain under perennial forage cover.
- Scenario 2 assumes that ΔrA will all be seeded to annuals, such as for food (bread quality wheat or cooking oil) or feedstock for grain ethanol or canola biodiesel.
- Scenario 3 assumes that ΔrA will be returned to beef production with the same overall herd structure and, hence, the same overall GHG emission rates per ha.
- Scenario 4 assumes that ΔrA will be returned to beef production with a mainly grass fed population whose diet included more forage and less grain than the beef cattle in Scenario 3.

Scenario 1 can be considered a baseline for the other three scenarios. Scenarios 2, 3 and 4 increase the carbon footprint of the beef to pork redistribution and had to be subtracted from the initial GHG emission savings.

Area based GHG emission rates for edible pulses and cereals defined by Dyer et al. [45] were used in Scenario 2 to account for growing annual crops on ΔrA. In Scenarios 3 and 4, the emissions to be subtracted from the initial GHG emission savings come from repopulating ΔrA with beef cattle. For Scenario 3, the emissions to be subtracted were based on the whole beef population. For Scenario 4, the emissions to be subtracted were based on just those parts of the beef population that were not being fattened for slaughter with feed grain supplements.

6.3 RESULTS AND DISCUSSION

6.3.1 RESULTS

6.3.1.1 CANADIAN LIVESTOCK GHG AND LAND USE INVENTORY

A summary of GHG emissions from the four livestock industries for eastern (Atlantic Provinces, Québec, Ontario) and western (Manitoba, Saskatchewan, Alberta, British Columbia) Canada is shown in Figure 2. These quantities reflect the respective sizes of the four industries as much as differences in GHG emission types. Since changes in soil carbon relate to interactions among livestock populations, GHG emissions were grouped in a way that most closely relates to ruminant and non-ruminant livestock systems. Hence, the GHG emissions in Figure 2 are distinguished as either enteric or non-enteric. Non-enteric GHG emissions include manure methane, N_2O from both the soil and stored manure, and fossil CO_2. The main sources of the non-enteric GHGs are the annual crops that supply the feed grains for non-ruminants (hogs and poultry), and the grain component of cattle diets. Canadian livestock accounted for 53 $TgCO_2e$ in 2001 with 22 $TgCO_2e$ coming from enteric methane. The Canadian beef industry emitted 31 $TgCO_2e$. Western beef accounted for 26 $TgCO_2e$, 14 of which were enteric methane. Dairy and pork production accounted for 10 and 7 $TgCO_2e$, respectively. At 5 $TgCO_2e$, poultry was the lowest source of GHG from the livestock industry in Canada.

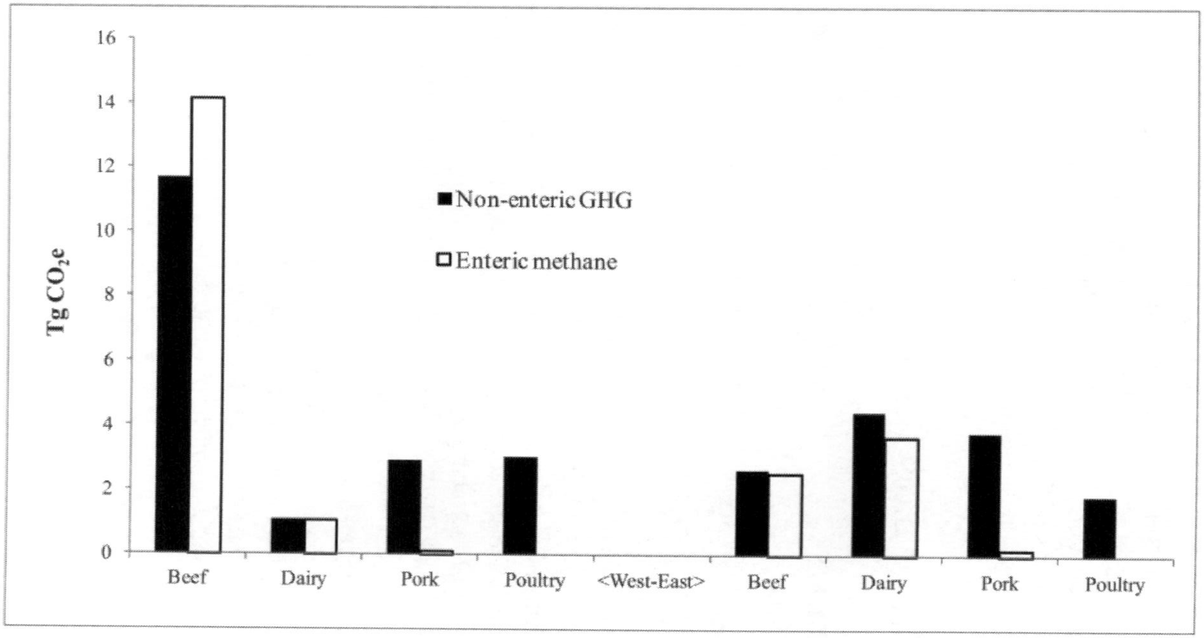

FIGURE 2: Greenhouse gas (GHG) emissions from four types of livestock in eastern and western Canada separated into enteric and non-enteric sources (land use and manure storage systems) in 2001.

Figure 3 shows the areas in each of these four crop complexes and groups land use according to those cultivated for grain, harvested forage, or improved pasture. The BCC and DCC include all three classes of land (grain, forage and pasture), but the only land in the PCC and ACC is land that can produce grains and pulses (annuals). Since soil carbon is generally higher under perennial forage than under annual crops [10], a shift from ruminant to non-ruminant livestock production would reduce soil carbon stock. Silage corn, although an annual crop, was grouped with the forages in Figure 3. Pasture represents an appreciable land use only in the western beef industry. It was assumed in this study that most of that land would be unsuitable, or at least the last land selected, for reseeding to grow annual feed grains or harvested field crops. Hence, that land would most likely continue to be under permanent (perennial) cover under the 10% livestock redistribution scenarios examined in this paper.

6.3.1.2 IMPACTS OF BEEF TO PORK REDISTRIBUTION ON CROP LANDS AND GHG EMISSIONS

In Table 1, the pork inflation factor is closer to 100% than the beef deflation factor (90%) in the east because the total supply of pork protein is much greater than the total supply of protein from beef in that region. The opposite is true in the west because the total supply of beef protein is higher than the total pork protein supply in that region. The post-redistribution areas in Table 2 reflect the percentages shown in Table 1, whereby the proportional increase in land resources allocated to pork production is smaller than the proportional decrease in areas supporting beef in the east, but is larger in the west. As expected, the new areas to produce feed grains for the expanded hog population exceeded the reduction in areas that had produced feed grains and silage corn for beef in both regions. Table 2 also shows that silage corn was the dominant annual crop for beef feed in the east, but was not important in the west. Even though there is 3.4 times as much land in these two industries in the west (Table 2), the total protein supplied by beef and pork from the west only exceeds the east by 80% (Table 1).

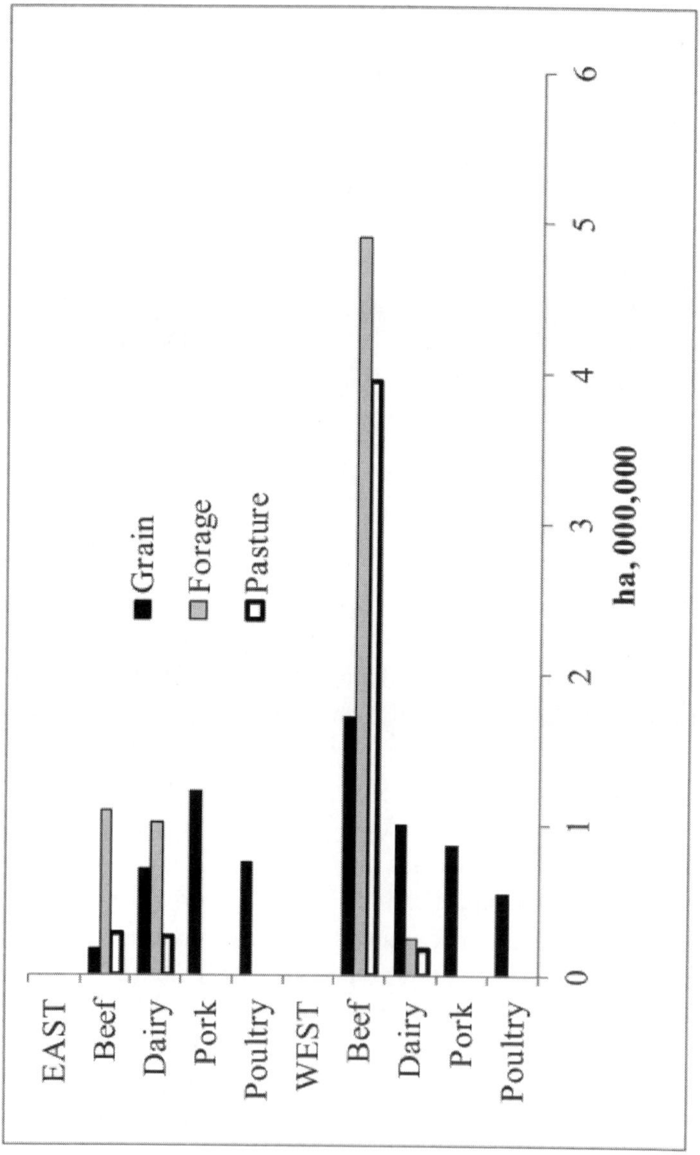

FIGURE 3: Areas in the livestock crop complex in each of four types of livestock in eastern and western Canada in 2001 grouped by three general land use classes.

TABLE 2: Crop areas that support the beef and pork industries before and after redistribution of land from beef to pork production.

	Beef			Hogs
	Feed grain	Silage corn	Harvested perennials	Feed grain
	Before redistribution (ha.10^3)			
East	117	98	1,000	1,219
West	1,818	42	4,944	1,641
Canada	1,994	140	5,944	2,860
	After redistribution (ha.10^3)			
East	159	88	900	1,247
West	1,636	38	4,449	1,932
Canada	1,795	126	5,350	3,179

The new area in annuals in Table 3 (first column, ΔcA) does not affect GHG emissions among the four scenarios because the emissions are already accounted for in the area of expanded pork. The areas in the second column, ΔcA + ΔrA, signify the initial reduction of the beef population, before reallocating ΔcA back to the pork industry. The third column, ΔrA, takes the areas required for the expanded pork production system shown in Column 1 into account. Column 3 is the result of the differences in areas in Table 2 for changes in both livestock populations. Because of the stronger role of silage corn in the eastern beef diet, the encroachment into perennial forage land area for grain production for hogs (Column 2) was much lower in the east. The portions of ΔrA that must be used to grow feed grain for the repopulated beef in the third and fourth scenarios are shown in the last two columns of Table 3. The areas for growing additional feed grain are about a third higher for Scenario 3 compared to Scenario 4. These relatively small portions of ΔrA reflect the lesser role of grains in the diet of the repopulated beef cattle.

TABLE 3: Changes in area in the beef crop complex as a result of reducing beef production and expanding pork production, and after repopulating the residual forage area (ΔrA) with beef cattle.

	New area in annuals (ΔcA) beef to pork	Area remaining from harvcested perennial forage		Annuals to support new beef	
		Initial[1]	Residual[2]	Mix of forage and grain	Mainly grass-fed
Regions			ha,000		
East	1.1	100.0	99.0	21.3	16.6
West	104.7	494.4	389.7	106.5	77.1
Canada	105.8	594.4	488.6	129.1	93.4

[1] *area freed after initial reduction in the beef cattle displaced by hogs (ΔcA + ΔrA)*
[2] *perennial area remaining after re-seeding to annuals to feed more hogs (ΔrA)*

The GHG emission budgets for beef and pork production prior to the four scenarios are presented in Table 4. The first group of four rows illustrates the basic livestock-specific GHG simulations generated by ULICEES. The western beef industry is by far the largest source of GHG emissions, and methane from western beef is the largest term in the combined GHG budget of these two industries. There was much less east-west difference in emissions in the pork industry, but both were lower than the eastern beef industry emissions. Fossil CO_2 was the lowest GHG emission from both industries while CH4 was the highest. The second and third groups of two rows represent GHG deducted from beef and added to pork production, respectively. The remaining data in Table 4 show the net potential savings or reductions in annual GHG emissions as a result of the beef to pork redistribution. The values in the fourth column and the last two rows represent the GHG emission changes from the redistribution from eastern and western Canada, respectively, prior to the scenario assessment. The first three quantities in the last two lines show that the beef to pork redistribution resulted in lower annual emissions for all three GHGs, but especially for methane because of the ruminant digestion of forage by cattle.

TABLE 4: Comparison of the annual GHG emission budgets of the 2001 Canadian beef and pork industries before and after land redistribution due to increased pork production.

		TgCO$_2$e			
Farm type	Region	CH$_4$	N$_2$O	CO$_2$	GHGs
		Baseline annual GHG emissions prior to land redistribution			
Beef	East	2.64	2.06	0.49	5.19
	West	14.71	8.32	2.78	25.81
Pork	East	1.62	1.44	0.92	3.99
	West	1.46	0.77	0.83	3.06
		Deducted GHG emissions resulting from reduced beef production			
Beef	East	0.26	0.21	0.05	0.52
	West	1.47	0.83	0.28	2.58
		Additional GHG emissions resulting from increased pork production			
Pork	East	0.04	0.03	0.02	0.09
	West	0.26	0.14	0.15	0.54
		Net annual GHG emissions deducted from land redistribution			
	East	0.23	0.17	0.03	0.43
	West	1.21	0.69	0.13	2.04

TABLE 5: Annual GHG emissions from the residual forage area (rA) under four land use scenarios 1 in eastern and western Canada in 2001.

Scenario #	1	2	3	4
	TgCO$_2$e			
East	0.00	0.19	0.40	0.50
West	0.00	0.29	1.48	1.81
Canada	0.00	0.48	1.88	2.31

1 four scenarios for ΔrA:
Scenario 1: remains under perennial forage cover,
Scenario 2: seeded to annuals,
Scenario 3: returned to beef production with mixed forage and grain diet,
Scenario 4: returned to beef production with mostly forage diet.

TABLE 6: Changes in soil carbon and annual GHG emissions due to beef to pork redistribution, and payback periods required for decreased GHG to compensate soil carbon losses, under four scenarios 1 for using residual land 2 in 2001.

Scenario #	1	2	3	4
	Soil carbon loss over 40 years (Tg CO2e)			
East	0.10	9.57	2.12	1.67
West	7.67	54.48	15.48	13.32
Canada	7.77	64.06	17.60	14.99
	Decrease 3 in annual GHG emissions (Tg CO2e)			
East	0.43	0.24	0.02	0.08
West	2.04	1.75	0.56	0.23
Canada	2.46	2.02	0.59	0.15
	Payback period (years)			
East	0.2	40.1	92.7	-
West	3.8	31.2	27.6	57
Canada	3.2	31.9	29.9	96.9

[1] *scenarios include four uses of residual land for crop or beef production*
[2] *residual land includes the area freed by displaced beef population*
[3] *negative quantities represent an increase in annual emissions*

The estimated annual GHG emissions from ΔrA under the four scenarios are shown in Table 5. These GHG emissions were subtracted from the GHG reductions in Table 4 to estimate the net changes in GHG emissions under each scenario. Due to enteric methane, the highest GHG emissions resulted from Scenario 4, followed by Scenario 3. The fertilizer N_2O and fossil CO_2 from farm field operations under Scenario 2 resulted in lower annual GHG emissions than from Scenarios 3 and 4. There were no GHG emissions from ΔrA under Scenario 1 since ΔrA is supposed to remain under perennial forage.

The expected 40 year losses in soil carbon [40] as a result of the four scenarios are shown Table 6 (rows 1 to 3). The net changes in annual GHG emissions associated with each of the four scenarios are then presented (rows 4 to 6). In Scenario 1, the net annual reduction in GHG emissions from Table 4 were compared to the changes in soil carbon under ΔcA.

For Scenarios 2, 3 and 4, the annual GHG emission changes had to be compared to changes in soil carbon under both ΔcA and ΔrA. The payback periods in years required for reductions in annual GHG emissions to equal the 40 year cumulative losses in the soil carbon stock were determined from the ratios of the 40 year soil carbon losses (rows 1 to 3) to the respective decreases in annual GHG emissions (rows 4 to 6).

The largest loss of soil carbon came from reseeding all of ΔrA to annual crops (Scenario 2, Table 6). The lowest loss in soil carbon results from leaving all of the residual land under perennial ground cover without repopulating ΔrA with beef cattle (Scenario 1). Repopulating ΔrA with just a category of beef that is highly forage dependent (Scenario 4) resulted in slightly less soil carbon loss than from repopulating ΔrA with beef cattle fed a mix of forage and grain (Scenario 3) in both eastern and western Canada. The net annual GHG emission reductions for Scenarios 3 and 4 are appreciably lower than the net annual emission reductions in Scenarios 1 and 2. The negative result for the eastern emission quantity for Scenario 4 indicates a net increase in annual GHG emissions. Because this scenario resulted in a loss of GHG mitigation potential in the east, there was no need to relate this annual loss to the loss of soil carbon.

Due to high methane emissions, payback periods were the highest for Scenario 4, even though the predominantly forage based cattle produced the second lowest decrease in soil carbon. The payback period for reseeding all residual perennial forage to annual crops (Scenario 2) was 13% longer than the mixed forage and grain fed beef herd (Scenario 3) in western Canada, but was only 40% as long as in the east. Due to the increase in annual GHG emissions in the east under Scenario 4, no payback period was shown for this case. For Scenario 3 in the east, the payback period was 2.3 times the 40 year benchmark period. In western Canada, the two ratios of payback to benchmark periods were 0.7 and 1.4 for Scenarios 3 and 4, respectively. For Scenario 2, the payback to benchmark period ratios were 1.0 and 0.8 for eastern and western Canada, respectively. For Scenario 1, the ratios were less than 0.1 for both eastern and western Canada. The very short payback periods for Scenario 1 were the result of the lowest change in soil carbon of any of the four Scenarios, as well as the unreduced GHG savings from the basic beef to pork redistribution.

6.3.2 DISCUSSION

From a purely GHG emissions perspective, the payback period of four years or less for Scenario 1 (Table 6) suggests that leaving ΔrA as unconsumed perennial growth was the best GHG mitigation option. While this benefit would require the elimination of livestock, Scenario 1 could be used to grow feedstock for cellulosic ethanol [46], or simply be set aside for environmental purposes like wildlife habitat. With payback periods at three quarters or less of the 40 year window, Scenario 2 represents a net gain in CO_2 mitigation potential in the west. In the east, Scenario 2 was neutral with respect to the 40 year payback window. While Scenario 2 eliminated livestock, the annual crops in this scenario would increase global food supply. Scenario 3 in the west was the only case where repopulating ΔrA with beef cattle led to a net gain in GHG mitigation potential based on the payback period being less than 40 years. Scenario 3 also had a slight advantage over Scenario 2 in the west. This was because beef production conserves soil carbon stock and it is a lower input system than field crops. Due to the heavy dependence on silage corn, this GHG mitigation benefit was lost in the eastern beef industry under Scenario 3. Scenario 4 was the least promising mitigation option, in spite of lower losses of soil carbon than under Scenario 3. This was particularly true in the east where the net annual GHG emissions actually increased.

The need to apply four scenarios to the basic beef to pork redistribution is consistent with previous studies involving the interaction between the beef industry and biofuel feedstock production [5,46]. Like the interaction with pork production in this study, these previous studies showed that displacement of beef with several types of biofuel feedstock has a range of outcomes depending on how the operators of the displaced beef farms respond. Similarly, whether or not the residual land in this analysis will be used sustainably depends largely on how much the beef farm operators will be allowed to share in the economic benefits of the land use shifts [46].

Because the beef to pork redistribution required land to be shifted from forage to annual grain crops, it was assumed that a sufficient portion (up to 10%) of the land under forage would be suitable for growing feed grains. Scenario 2 required that all of ΔrA be suitable for either cereal or pulse crop production. The main uncertainty in the beef to pork redistribution was the use of ΔrA. This was because that land may be used for either perennial or annual crops, or to support either food or biofuel feedstock production. It might also be allowed to revert back to rangeland or natural habitat, depending on the economic pressures. Hence, the four scenarios represent only a few of the many possible ways that livestock industries and soil carbon stocks can interact. When the GHG emissions from the Canadian livestock industries were assessed individually, the LCC approach allowed them to be treated as closed systems. However, once farm type interactions are introduced, inter-commodity land use changes can no longer be viewed in complete isolation because the different intensities of land do not result in equal exchanges of land areas.

Exploring scenarios for the redistribution shifted the focus from direct impacts on ΔcA to indirect impacts on ΔrA. However, it was not the objective of this paper to determine whether or not repopulating ΔrA with beef was necessarily the best use of the residual crop land. Because these scenarios were only applicable to ΔrA, this assessment does not describe the carbon footprint of the entire production system associated with each of the four scenarios. For example, the annual field crops in Scenario 2 by themselves, do not replace the protein that could have been produced if either Scenario 3 or 4 had been the chosen land use option. Even though these last two scenarios have not been found to mitigate GHG emissions, both of them resulted in higher beef protein production. Also, there are recognized ancillary environmental benefits to beef production that enhance sustainability [43], but this broader perspective on the overall sustainability of this production system was beyond the scope of this paper. Therefore, this scenario assessment should be treated as a demonstration of how the annual GHG emission budgets and soil carbon interact under changes in land use.

6.4 CONCLUSIONS

The model presented in this paper, was developed in order to quantify the impacts of changes in annual GHG emissions and soil carbon on the carbon footprint of the Canadian livestock industry. The paper also presents a method of reallocating farm animals or land from ruminant to non-ruminant production systems. With the inclusion of changes in soil carbon storage, the ULICEES model can provide a more comprehensive comparison of the negative impact of enteric methane emissions with the benefit of protecting soil carbon under permanent cover. Similarly, other livestock commodities or non-enteric related GHG emission issues, such as manure management systems or tillage practices, can also be compared with changes in soil carbon with ULICEES. With the soil carbon interface, ULICEES can provide a more comprehensive comparison of GHG emission intensities of protein production among Canadian livestock types.

Although it introduces an additional parameter to the GHG mitigation policy dialogue, the payback period approach was shown to be a potentially valuable indicator. For example, whereas reseeding all of the residual areas from displaced beef to annuals released the largest amounts of soil carbon, the savings in annual GHG emissions showed that this option could be a positive GHG emissions mitigation strategy over 40 years. However, given the vast reserve of grassland in western Canada that cannot support grain production, the lower carbon footprint of non-ruminants cannot override the fact that this land is only suitable for forage based beef production.

This study has illustrated the complex interactions between livestock production industries that must be considered when attempting to balance agricultural production, land use and mitigation strategies. ULICEES shows promise as being an effective modeling tool for a wide range of land use and GHG mitigation policies in Canada. The results of the scenarios assessed in this paper suggest that conserving soil carbon stock did not compensate for the annual GHG emissions from forage based beef production in much of Canada. These findings should not, however, be interpreted as an indication that all farmland should be converted from beef to pork production. Assessments similar to this Canadian analysis could be done in other countries. However, in countries whose lack of food security would not justify grain based livestock production, or where most of the land resources are only suited to perennial forage production, ruminant livestock would continue to be the most sustainable, and often the only viable, food production system, regardless of the carbon footprint of that system.

REFERENCES

1. Report of the Special Rapporteur on the Right to Food; Item 72(b) of the Provisional Agenda; Sixty Second Session; A/62/289; United Nations: New York, NY, USA, 22 August 2007; pp. 1–23.
2. Vergé, X.P.C.; Worth, D.E.; Dyer, J.A.; Desjardins, R.L.; McConkey, B.G. LCA of Animal Production. In Green Technologies in Food Production and Processing; Food Engineering Series; Chapter 5; Arcand, Y., Boye, J., Eds.; Springer Science+Business Media: New York, NY, USA, 2012; pp. 83–113.
3. Trivedi, B. What is your dinner doing to the climate? New Scientist magazine, 2673:28-32. Available online: http://www.mbayaq.org/cr/SeafoodWatch.asp. (Accessed on 11 September 2008).
4. Vergé, X.P.C.; Dyer, J.A.; Desjardins, R.L.; Worth, D. Greenhouse gas emissions from the Canadian beef industry. Agric. Syst. 2008, 98, 126–134.
5. Dyer, J.A.; Vergé, X.P.C.; Desjardins, R.L.; McConkey, B.G. Implications of biofuel feedstock crops for the livestock feed industry in Canada. In Environmental Impact of Biofuels; Chapter 9; Dos Santos Bernardes, M.A., Ed.; InTech: Rijeka, Croatia, 2011; pp. 161–178.
6. Dyer, J.A.; Vergé, X.P.C.; Kulshreshtha, S.N.; Desjardins, R.L.; McConkey, B.G. Residual crop areas and greenhouse gas emissions from feed and fodder crops that were not used in Canadian livestock production in 2001. J. Sustain. Agric. 2011, 35, 780–803.
7. Storm, I.M.L.D.; Hellwing, A.L.F.; Nielson, N.I.; Madsoen, J. Methods for measuring and estimating methane emissions from ruminants. Animals 2012, 2, 160–183.
8. Flachowsky, G.; Kampheus, J. Carbon footprints for food of animal origin: What are the most preferable criteria to measure animal yields? Animals 2012, 2, 108–126.
9. Boehm, M.; Junkins, B.; Desjardins, R.; Kulshreshtha, S.; Lindwall, W. Sink potential of Canadian agricultural soils. Clim. Chang. 2004, 65, 297–314.
10. Janzen, H.H.; Desjardins, R.L.; Rochette, P.; Boehm, M.; Worth, D. Better Farming Better Air—A Scientific Analysis of Farming Practices and Greenhouse Gases in Canada; Agriculture and Agri-Food Canada: Ottawa, ON, Canada, 2008; p.146.
11. Desjardins, R.L.; Vergé, X.; Hutchinson, J.J.; Smith, W.N.; Grant, B.; McConkey, B.; Worth, D. Greenhouse Gases; Lefebvre, A., Eilers, W., Chun, B., Eds.; Agri-Environmental Indicators Report 2005; Agriculture and Agri-Food Canada: Ottawa, ON, Canada, 2005; pp. 142–148.
12. Janzen, H.H.; Angers, D.A.; Boehm, M.; Bolinder, M.; Desjardins, R.L.; Dyer, J.; Ellert, B.H.; Gibb, D.J.; Gregorich, E.G.; Helgason, B.L.; et al. A proposed approach to estimate and reduce net greenhouse gas emissions from whole farms. Can. J. Soil Sci. 2006, 86, 401–418.
13. Dyer, J.A.; Vergé, X.; Desjardins, R.L.; Worth, D. Long term trends in the GHG emissions from the Canadian Dairy industry. Can. J. Soil Sci. 2008, 88, 629–639.
14. Vergé, X.P.C.; Dyer, J.A.; Desjardins, R.L.; Worth, D. Greenhouse gas emissions from the Canadian dairy industry during 2001. Agric. Syst. 2007, 94, 683–693.
15. Vergé, X.P.C.; Dyer, J.A.; Desjardins, R.L.; Worth, D. Greenhouse gas emissions from the Canadian pork industry. Livest. Sci. 2009, 121, 92–101.

16. Vergé, X.P.C.; Dyer, J.A.; Desjardins, R.L.; Worth, D. Long Term trends in greenhouse gas emissions from the Canadian poultry industry. J. Appl. Poult. Res. 2009, 18, 210–222.
17. Klein, K.K.; LeRoy, D.G. The Biofuels Frenzy: What's in it for Canadian Agriculture? Green Paper Prepared for the Alberta Institute of Agrologists. In Proceedings of the Annual Conference of Alberta Institute of Agrologists, Banff, Alberta, Canada, 28 March 2007; p. 46.
18. Elward, M.; McLaughlin, B.; Alain, B. Livestock Feed Requirements Study 1999–2001; Catalogue No. 23-501-XIE; Statistics Canada: Ottawa, ON, Canada, 2003; p. 84.
19. Fournel, S.; Pelletier, F.; Godbout, S.; Legace, R.; Feddes, J. Greenhouse gas emissions from three layer housing systems. Animals 2012, 2, 1–15.
20. Cattle Statistics; 23-012-XIE; Statistics Canada: Ottawa, ON, Canada, 2005.
21. Hog Statistics; 23-011-XIE; Statistics Canada: Ottawa, ON, Canada, 2005.
22. Intergovernmental Panel on Climate Change (IPCC). Intergovernmental Panel on Climate Change: Guidelines for national greenhouse gas inventories. Agric. For. Other Land Use 2006, 4, 87.
23. Vergé, X.; Worth, D.; Hutchinson, J.; Desjardins, R. Greenhouse Gas Emissions from Canadian Agro Ecosystems—Technical Report; A22-414/2006E; Agriculture and Agri-Food Canada: Ottawa, ON, Canada, 2006; p. 38.
24. Jørgensen, H. Methane emission by growing pigs and adult sows as influenced by fermentation. Livest. Sci. 2007, 109, 216–219.
25. Intergovernmental Panel on Climate Change (IPCC). Good Practice Guidance and Uncertainty in National Greenhouse Gas Inventories. Section 4: Agriculture. 2000. Available online: http://www.ipcc-nggip.iges.or.jp/public/gp/gpg-bgp.htm (accessed on 27 August 2012).
26. Rochette, P.; Worth, D.E.; Lemke, R.L.; McConkey, B.G.; Pennock, D.J.; Wagner Riddle, C.; Desjardins, R.L. Estimation of N2O emissions from agricultural soils in Canada—Development of a country specific methodology. Can. J. Soil Sci. 2008, 88, 641–654.
27. Yang, J.Y.; de Jong, R.; Drury, C.F.; Huffman, E.C.; Kirkwood, V.; Yang, X.M. Development of a Canadian agricultural nitrogen budget (CANB v2.0) model and the evaluation of various policy scenarios. Can. J. Soil Sci. 2007, 87, 153–165.
28. Korol, M. Canadian Fertilizer Consumption, Shipments and Trade 2001/2002. Farm Input Market Unit, Agriculture and Agri-Food Canada, 2002. Available online: http://www.cfi.ca/_documents/uploads/elibrary/cf01_02_e[1].pdf (accessed on 27 August 2012).
29. Dyer, J.A.; Desjardins, R.L. Energy based GHG emissions from Canadian Agriculture. J. Energy Inst. 2007, 80, 93–95.
30. Dyer, J.A.; Desjardins, R.L. A review and evaluation of fossil energy and carbon dioxide emissions in Canadian agriculture. J. Sustain. Agric. 2009, 33, 210–228.
31. Dyer, J.A.; Desjardins, R.L. Simulated farm fieldwork, energy consumption and related greenhouse gas emissions in Canada. Biosyst. Eng. 2003, 85, 503–513.
32. Dyer, J.A.; Desjardins, R.L. Analysis of trends in CO2 emissions from fossil fuel use for farm fieldwork related to harvesting annual crops and hay, changing tillage practices and reduced summerfallow in Canada. J. Sustain. Agric. 2005, 25, 141–156.

33. Dyer, J.A.; Desjardins, R.L. Carbon dioxide emissions associated with the manufacturing of tractors and farm machinery in Canada. Biosyst. Eng. 2006, 93, 107–118.
34. Dyer, J.A.; Desjardins, R.L. An integrated index for electrical energy use in Canadian agriculture with implications for greenhouse gas emissions. Biosyst. Eng. 2006, 95, 449–460.
35. James, A.; Brown, P. Special report: Global warming, Carbon dioxide levels will double by 2050, experts forecast. The Guardian (Environment), 6 April 2001. Available online: http://www.guardian.co.uk/environment/2001/apr/06/usnews.globalwarming (accessed on 15 August 2012).
36. Environmental Protection Agency (EPA). Accounting Framework for Biogenic CO2 Emissions from Stationary Sources. Office of Atmospheric Programs, Climate Change Division; United States Environmental Protection Agency (EPA): Washington, DC, USA, 2011. Available online: http://www.epa.gov/climatechange/emissions/biogenic_emissions.html (accessed 15 August 2012).
37. Fargione, J.; Hill, J.; Tilman, D.; Polasky, S.; Hawthorne, P. Land clearing and the biofuel carbon debt. Science 2008, 319, 1235–1238.
38. Marine Corps Community Services (MCCS). Massachusatts Biomass Sustainability and Carbon Policy Study; Report to the Commonwealth of Massachusatts Department of Energy Resources; NCI-2010-03; Manomet Center for Conservation Sciences (MCCS): Brunswick, ME, USA, 2010. Available online: http://www.manomet.org/sites/manomet.org/files/Manomet_Biomass_Report_Full_LoRez.pdf (accessed on 27 August 2012).
39. Searchinger, T.; Heimlich, R.; Houghton R.A.; Dong, F.; Elobeid, A.; Fabiosa, J.; Tokgoz, S.; Hayes, D.; Yu, T.-H. Use of U.S. croplands for biofuels increases greenhouse gases through emissions from land use change. Science 2008, 319, 1238–1240.
40. McConkey, B.G.; Anger, D.A.; Bentham, M.; Boehm, M.; Brierley, A.; Cerkoniak, D.; Liang, B.C.; Collas, P.; de Gooijer, H.; Desjardins, R.L.; et al. Canadian Agricultural Greenhouse gas Monitoring Accounting and Reporting System. Methodology and Greenhouse Gas Estimates for Agricultural Land in the LULUCF Sector for NIR 2006; Report submitted to the Greenhouse Gas Division, Environment Canada; Research Branch of Agriculture and Agri-Food Canada: Ottawa, ON, Canada, 2007; p. 105.
41. Dyer, J.A.; Vergé, X.P.C.; Desjardins, R.L.; Worth, D.E. The Protein based GHG emission intensity for livestock products in Canada. J. Sustain. Agric. 2010, 34, 618–629.
42. Capper, J.L. The environmental impact of beef production in the United States: 1977 compared with 2007. J. Anim. Sci. 2011, 89, 4249–4261.
43. Beauchemin, K.A.; Janzen, H.H.; Little, S.M.; McAllister, T.A.; McGinn, S.M. Life cycle assessment of greenhouse gas emissions from beef production in western Canada: A case study. Agric. Syst. 2010, 103, 371–379.
44. Basarab, J.; Baron, V.; López-Campos, O.; Aalhus, J.; Haugen-Kozyra, K.; Okine, O. Greenhouse gas emissions from Calf- and yearling-fed beef production systems, with and without the use of growth promotants. Animals 2012, 2, 195–220.
45. Dyer, J.A.; Vergé, X.P.C.; Desjardins, R.L.; Worth, D.E.; McConkey, B.G. The impact of increased biodiesel production on the greenhouse gas emissions from field crops in Canada. Energy Sustain. Dev. 2010, 14, 73–82.

46. Dyer, J.A.; Hendrickson, O.Q.; Desjardins, R.L.; Andrachuk, H.L. An environmental impact assessment of biofuel feedstock production on agro-ecosystem biodiversity in Canada. In Agricultural Policies: New Developments. Chapter 3:87-115; Contreras, L.M., Ed.; Nova Science Publishers Inc.: Hauppauge, NY, USA, 2011; p. 281.

CHAPTER 7

DIESEL CONSUMPTION OF AGRICULTURE IN CHINA

NAN LI , HAILIN MU, HUANAN LI, AND SHUSEN GUI

7.1 INTRODUCTION

The limited supply of traditional fossil fuels and the associated consumption limitations needed for dealing with global climate change have considerably restricted economic development in recent years. This has raised the concern of energy analysts and policy makers regarding the adverse effects of energy overuse. A series of policy measures aimed at reducing energy consumption have been implemented in order to meet the compulsory targets stated in the China government's Eleventh Five-Year Plan (2006–2010). Agriculture development in China is not an exception. Calculated at constant prices, the average elasticity of energy consumption in agriculture declined from 2.33 (2001–2005) to 0.29 (2006–2010). The rapid growth of energy consumption in agriculture was thus restrained without negatively affecting economic development.

With a gradually increasing level of mechanization, agricultural energy consumption in China has increased from 36.88 million tons of coal equivalent (Mtce) in 1996 to 64.77 Mtce in 2010, which translates to an annual increase of 4.10%. Energy plays a critical role in the development

This chapter was originally published under the Creative Commons Attribution License. Li N, Mu H, Li H, and Gui S. Diesel Consumption of Agriculture in China. Energies ***2012**,5 (2012): 5126-5149. doi:10.3390/en5125126.*

of agriculture as it does in the manufacturing, construction and service industry. This has motivated many researchers to focus on agricultural energy issues, and analysis of energy and exergy efficiency in the agricultural sector has become a research hotspot. Two energy resources, namely diesel for tractors and electricity for pumps are usually the research topics in this area. Such analysis has been applied in Saudi Arabia [1], Turkey [2], Jordan [3], Iran [4] and Malaysia [5]. The relationship between energy inputs and agricultural production outputs is another research hotspot. These energy inputs usually include direct and indirect energy, i.e., human and animal labor, machinery, electricity, diesel oil, fertilizers, seeds, etc. Rijal and Bansal [6] examined the total energy input and output of subsistence agriculture in the rural areas of Nepal. Ozkan and Akcaoz [7] estimated the input-output ratio in the Turkish agricultural sector for the period of 1975–2000, where their output is composed of 36 agricultural commodities. On the other hand, agricultural output in Hatirli and Ozkan [8] comprises 104 agricultural commodities. Alam and Alam [9] evaluated the impact of energy input on agricultural production output in Bangladesh from 1980 to 2000. There are also researches that provided meaningful econometrics methods. Uri [10] quantified the relationship between the energy price and the use of conservation tillage via Granger causality over the period of 1963–1997. A regression analysis of the relationship between energy use and agricultural productivity was done by Karkacier and Gokalp Goktolga [11]. Using co-integration and error correction analysis, Türkekul and Unakıtan [12] estimated the long- and short-run relationship among energy consumption, agricultural GDP, and energy prices from 1970 to 2008 in Turkey's agriculture.

Based on a bottom-up modeling approach, the model named Save Production simulated the development of energy use in the Dutch industry and agriculture [13]. Baruah and Bora [14] assessed the energy demands in the state of Assam, India. In that study, they considered four strategic scenarios of mechanization that incorporated some proven technologies. Nevertheless, only a few simulation and forecast models were established to study the energy demand of agriculture and little attention has been paid to the relationship between energy consumption and end-use machinery in agriculture, especially in China. In the Twelfth Five-Year Plan, the Chinese government has planned to reduce the energy consumption per unit GDP by 16% during this five year horizon. A special model, Simulation and Analysis of Energy Consumption for Agriculture (SECA), is designed to answer all kinds of questions on how energy was consumed in different agricultural sectors to achieve the new goal. Furthermore it also serves as the foundation for the agricultural energy demand forecasting model. In this study, we identify the factors (total power, unit diesel consumption, etc.) influencing diesel consumption in China and simulate the diesel flows of agriculture of China in detail. Based on availability, our dataset spans the period of 1996–2010.

7.2 MODEL SPECIFICATION AND DATA

7.2.1 OVERALL STRUCTURE

This paper uses a generalized definition of the word agriculture to include farming (i.e., agriculture in narrow definition), fishery, forestry, animal husbandry and services supporting agriculture. Fifteen kinds of the agricultural machinery from seven agricultural subsectors are considered. Figure 1 shows the overall structure of SECA. In the Distribution Module, it is assumed that the change in the number of agricultural machinery with different rated powers follows a certain curved distribution. Simulation results of the curved distribution are processed by the Operation Module and transferred into the Unit Consumption Module and the Productivity Module which generate the weighted average of unit consumption and that of unit productivity, respectively. Then the two weighted averages are input into the Main Module along with workload statistics, machinery capacity and load factors, which are obtained from the database in the Operation Module. Finally, technological progress and other effects are considered in the Correction Module to narrow the gap between the empirical statistics and the model calculations. The following subsections describe each module in detail.

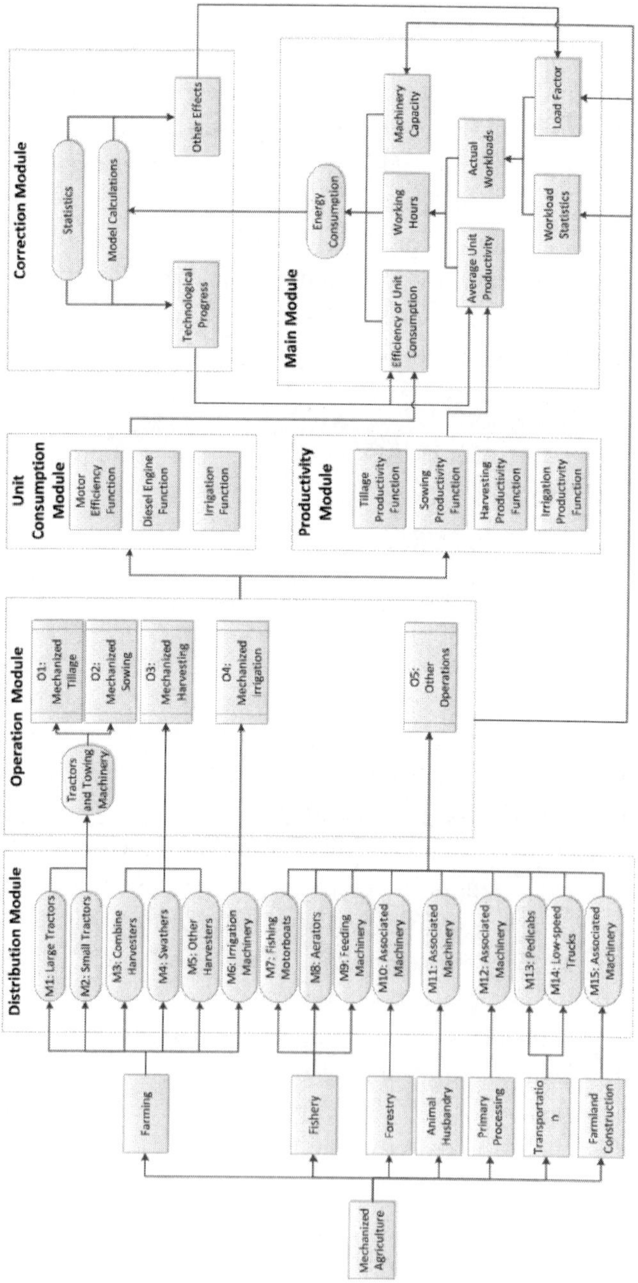

FIGURE 1: Overall structure of SECA.

TABLE 1: The relationships between operations and machinery

Operations	Machinery	Relationships
o1: Mechanized tillage	m1: Large and medium-sized tractors m2: Small tractors (and associated towing machinery)	$T_{01,y} = \sum_{m=m1}^{m2} T_{m,y}$
o2: Mechanized sowing	m1: Large and medium-sized tractors m2: Small tractors (and associated towing machinery)	$T_{02,y} = \sum_{m=m1}^{m2} T_{m,y}$
o3: Mechanized harvesting	m3: Combine harvesters m4: Swathers m5: Other harvesters	$T_{03,y} = \sum_{m=m3}^{m5} T_{m,y}$
o4: Mechanized irrigation	m6: Irrigation machinery	$T_{04,y} \sum_{m=m6}^{m6} T_{m,y}$
o5: Other operation	m7~m15: Rest machinery	$T_{05,y} \sum_{m=m7}^{m15} T_{m,y}$

Note: T is a generalized symbol which can be replaced by P, L, li, C, etc.

7.2.2 OPERATION MODULE

In SECA, all the agricultural machinery are categorized into five farming related operations (o1~o5, Table 1). These operations include mechanized tillage, mechanized sowing, mechanized harvesting, mechanized irrigation and other operations. They are assumed as follows:

- Mechanized tillage (o1) and mechanized sowing (o2) are performed by tractors and associated towing farm machinery. Both the large and medium-sized tractors and the small ones are included.
- Mechanized harvesting (o3) is performed by combine harvesters, swathers and other harvesters.

- Mechanized irrigation (o4) is performed by irrigation machinery.
- Other operations (o5) are performed by rest agricultural machinery, the workload statistics of which are not available.

7.2.3 MAIN MODULE

Two kinds of the energy carriers, diesel and electricity, are considered in SECA. The energy required to perform the selected farming operation is estimated using the equation below:

$$C_y = \sum_{o} C_{o,y} = \sum_{0} P_{o,y} \times t_{o,y} \times \overline{uc}_{o,y} \tag{1}$$

wherein the subscript o indicates the farm operations mentioned in the Operation Module (o = o1, o2,…, o5); the subscript y indicates the year (y = 1996, 1997,…, 2010); C refers to the quantity of energy consumption; P is the total power of agricultural machinery; t refers to the average annual working hours of unit machinery; uc is the weighted average of unit energy consumption (for diesel) or motor efficiency (for electricity).

The workload statistics for the four categories of agricultural operations, viz. o1, o2 , o3 and o4 , can be obtained from specific statistics. Their average annual working hours of unit machinery $t_{o,y}$ can be determined from the following equation:

$$t_{o,y} = \frac{AW_{o,y}}{\overline{up}_{o,y}} = \frac{WS_{o,y} \times lf_{o,y} \times oec_y}{\overline{up}_{o,y}} \tag{2}$$

wherein AW is the actual workloads of the farming operation; up indicates the weighted average of unit productivity; WS refers to the workload statistics; lf is the load factor, which is used for describing the actual work intensity and is assumed to be proportional to the machinery power in unit area; oec is the other effect coefficient obtained from the Correction Module.

Because workload statistics are not available for "other operations" (o5), in this study we assume that its average annual working hours of unit machinery t_{o5} is constant. That constant is determined from the following equation:

$$t_{o5} = \frac{\sum_y (P_{o5,y} \times \overline{uc}_{o5,y} \times \Delta C_y)}{\sum_y (P_{o5,y} \times \overline{uc}_{o5,y})^2} = \frac{\sum_y (P_{o5,y} \times \overline{uc}_{o5,y}) \times (c_{s,y} - \sum_{0=01}^{o4} C_{o,y})}{\sum_y (P_{o5,y} \times \overline{uc}_{o5,y})^2}$$

(3)

wherein $C_{s,\,y}$ indicates the diesel consumption statistics in year y .

7.2.4 DISTRIBUTION MODULE

The mechanization of agriculture is a process of replacing human and animal with agricultural machinery powered by either diesel or electricity. The rated power is an important factor affecting the device performance of agricultural machinery. The rated power of the agricultural machinery is negatively correlated with its unit consumption (or motor efficiency) and is positively correlated with its unit productivity.

Smaller agricultural machinery are more widely used than larger equipment in China due to reasons such as the high cost of larger machinery and the small amount of arable land per capita. It is assumed that the change in the number $l_{i,m,y}$ of agricultural machinery (except tractors) with the representative power $p_{i,m,y}$ of the interval i follows an exponential distribution as shown in Figure 2(a). The equation is expressed below:

$$l_{i,my} = a_{m,y}\ x\ e^{b,m,y\ x\ pi,m,y} \quad (4)$$

wherein m indicates machinery category mentioned in the Distribution Module (m = m3,m4,...,m15); $a_{m,y}$ and $b_{m,y}$ are the undetermined parameters.

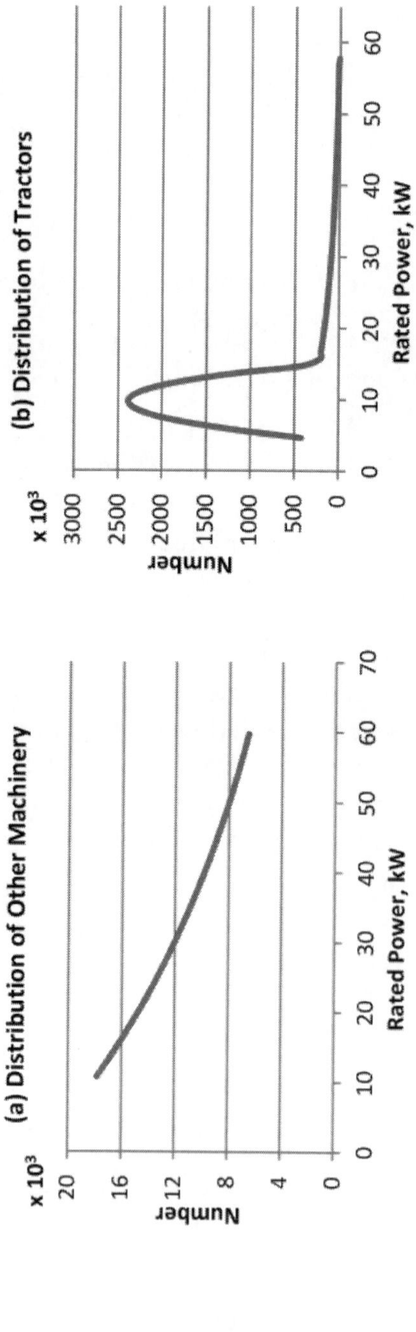

FIGURE 2: Distribution curves of the agricultural machinery with rated power.

In the Distribution Module, the most common power range ($P_{beginning,m,y}$, $P_{end,m,y}$) of machinery is selected on the basis of experience and divided into n intervals. The midpoint of interval i as its representative power $p_{i,m,y}$ can be obtained using the following relationship:

$$P_{i,m,y} = P_{beginning,m,y} + \frac{(2i-1)\times(P_{end,m,y} - P_{beginning,m,y})}{2n} \tag{5}$$

The total number of machinery $L_{m,y,c}$ and the total power $P_{m,y,c}$ can be obtained from statistics, so another two equations can be established:

$$L_{m,y} = \sum_i l_{i,m,y} \tag{6}$$

$$P_{m,y} = \sum_i (P_{i,m,y}\times l_{i,m,y}) \tag{7}$$

Equations (4)–(7) should be solved simultaneously to estimate parameters $a_{m,y}$ and $b_{m,y}$. Then the number $l_{i,m,y}$ of the agricultural machinery (excluding tractors) with the representative power $p_{i,m,y}$ in the interval i can be determined.

Tractors are the most common agricultural machinery and play an important role in mechanized agriculture in China. The number and total power of tractors increased from 9.86 million and 1.08×10^8 kW in 1996 to 21.78 million and 2.84×10^8 kW in 2010, with annual increases of 5.82% and 7.16%, respectively. The rated power of the main tractor models in Chinese market is between 8 and 12 kW, while in China's Department of Statistics, the rated power for tractors starts from 2.2 kW, so in SECA, it is assumed that the number of large and medium-sized tractors, referring to tractors with power ratings greater than or equal to 14.7 kW, still follows an exponential distribution. On the other hand, the number of small tractors, referring to tractors less than 14.7 kW and greater than 2.2 kW, follows a parabolic distribution [Figure 2(b)]:

$$l_{i,m1,y} = a_{m1,y} \times e^{b_{m1,y} \times P_{i,m1,y}} \quad (8)$$

$$l_{i,m2,y} = a_{m2,y} \times \left(P_{i,m2,y} - b_{m2,y}\right)^2 + c_{m2,y} \quad (9)$$

wherein $c_{m,y}$ is the undetermined parameter.

Another equation is established as follow in order to ensure the continuity of Equations (8) and (9):

$$l_{n+1,m2,y} = l_{1,m1,y} \quad (10)$$

Equations (5)–(10) should be solved simultaneously to estimate parameter $a_{m,y}$, $b_{m,y}$ and $c_{m,y}$. Then the number of tractors with the representative power in the interval i can be determined.

7.2.5 UNIT CONSUMPTION MODULE AND PRODUCTIVITY MODULE

Unit consumption (or motor efficiency) and productivity of agricultural machinery are both functions of the machinery's rated power. The simulation results from the Distribution Module are processed by the Operation Module and then output into the Unit Consumption Module and the Productivity Module which generate the weighted average of unit consumption and that of productivity. The data describing the relationship between the input and output of the functions are collected from the relevant national standards of China and product manuals of those agricultural machinery [15,16]. They are fitted by the least square method (Figures 3 and 4).

The fitting equation of the unit consumption is as follows:

$$u_{i,o,y} = \begin{cases} 0.0082 \times p_{i,o,y}^2 - 2.0807 \times p_{i,o,y} + 398.21 \; (diesel; o \neq o4) \\ 534.32 \times p_{i,o,y}^{-0.136} \; (diesel; o \neq o4) \\ 3.6572 \times \ln(p_{i,o,y}) + 77.516 \; (electricity) \end{cases} \quad (11)$$

The unit productivity for each operation is estimated using the following equations:

$$up_{i,o,y} = \begin{cases} 0.9903 \times p_{i,o1,y}^2 + 26.72 \times p_{1,o1,y} + 379.12 \; (o = o1) \\ 2068.6 \times e^{0.0235 \times p_{1,o2,y}} \; (o = o2) \\ 0.1875 \times p_{i,o3,y}^2 + 71.6 \times P_{i,o3,y} + 386.6 \; (o = o3) \\ 8.9867 \times p_{i,o4,y} + 10.23 \; (o = o4) \end{cases} \quad (12)$$

As the output in the Unit Consumption Module and the Productivity, the weighted averages of unit consumption and productivity are obtained according to the following relationships:

$$\overline{uc}_{o,y} = \frac{\sum_i (l_{i,o,y} \times uc_{i,o,y})}{L_{o,y}} \times tpc_{uc,y} \quad (13)$$

$$\overline{up}_{o,y} = \frac{\sum_i (l_{i,o,y} \times up_{i,o,y})}{L_{o,y}} \times tpc_{up,y} \quad (14)$$

wherein tpc_{uc} and tpc_{up} are the technological progress coefficients obtained from the Correction Module.

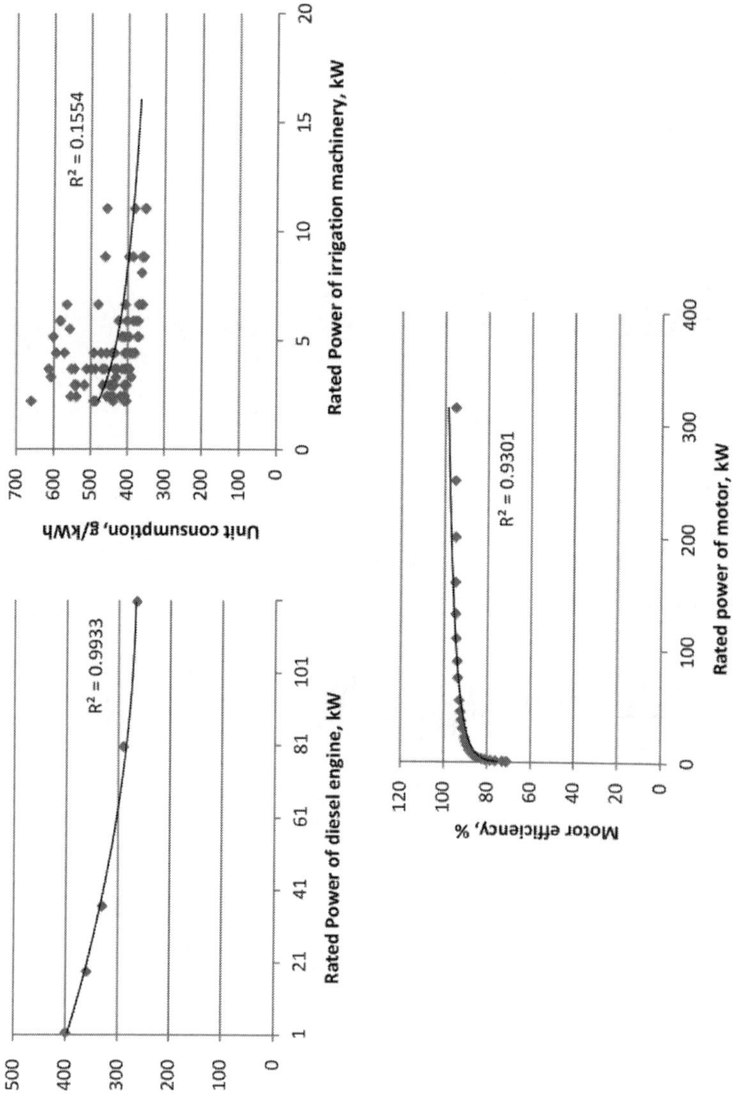

FIGURE 3: Fitting curves of the unit consumption.

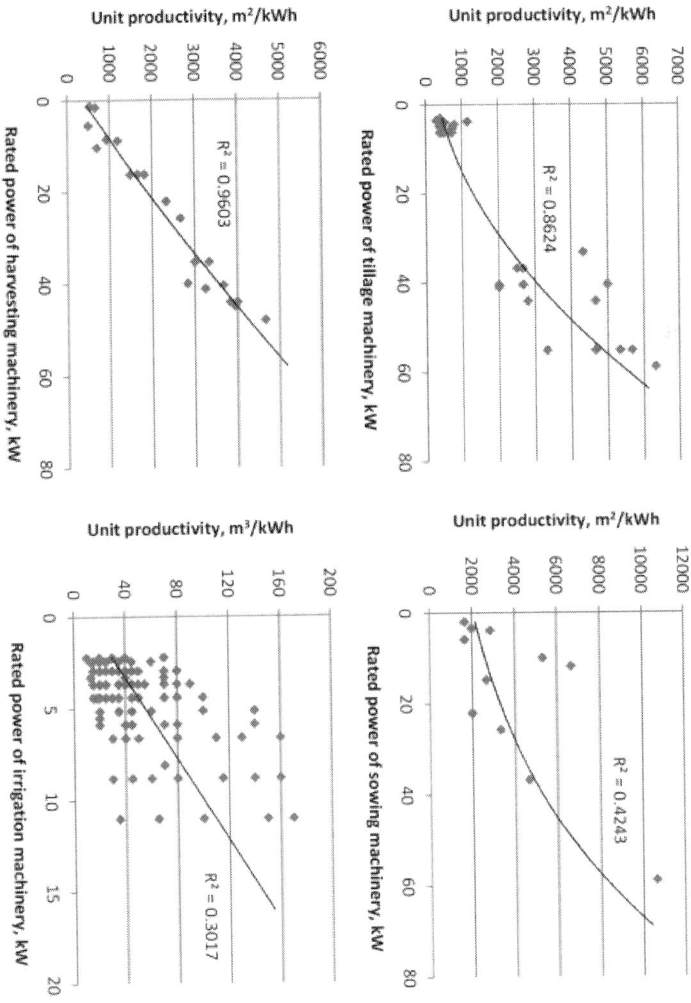

FIGURE 4: Fitting curves of the unit productivity.

7.2.6 CORRECTION MODULE

Technological progress and other effects are considered in the Correction Module to narrow the gap between the statistics and the model calculations in the Main Module. The correction coefficient is the correction on the basic assumptions in the other modules of SECA. These assumptions which are either explicit or implicit include:

- In the Unit Consumption Module, the weighted average of unit consumption changes without being affected by the technological progress.
- In the Productivity Module, the weighted average of unit productivity changes without being affected by the technological progress.
- In the Main Module, the load factor is proportional to the machinery power in unit area.
- In the Main Module, the average annual working hours of the other operation is constant the value of which remains unchanged over the year.

The correction coefficient coc can be calculated with the following relationship:

$$coc_y = \frac{C_{s,y}}{C_y} \tag{15}$$

The correction coefficient coc can be divided into two parts: technological progress coefficient tpc and the other effect coefficient oec. The relationship between coc, tpc and oec is assumed to following equation:

$$tpc_y = oec_y = coc_y^{1/2} \tag{16}$$

Technological progress is typically accompanied with the reduction of the unit consumption and the increase of the unit productivity. The relationship between tpc, tpc_{uc} and tpc_{up} is assumed according to Equation (17):

$$tpc_{uc,y} = tpc_{up,y} = tpc_y^{1/2} \tag{17}$$

The other effect coefficient oec mainly works on the load factor lf mentioned in the Main Module. The other effect coefficient oec means that the load factor is not proportional to machinery power in unit area any longer. In short, we can conclude that technological progress leads to the reduction of average unit consumption and changes in working hours result from the changes in unit productivity and load factor.

7.3 DATA SOURCES

The data related to the agricultural land, machinery and energy consumption in this paper are mainly obtained from the China Statistical Yearbook [17], China Energy Statistical Yearbook [18], China Rural Statistical Yearbook [19] and China Agriculture Statistical Report [20]. The first three data resources are published by the National Bureau of Statistics of China and the last one is published by the Ministry of Agriculture of China.

The data related to land and water resources in this paper are given by the China Land and Resources Bulletin [21] and the China Water Resources Bulletin [22]. The former is published by the Ministry of Land and Resources of China, and its data, especially the farmland area data, are more reliable than the other data sources [23]. The latter is published by the Ministry of Water Resources of China.

Table A1 in the Appendix presents the workload statistics of four operations: i.e., mechanized tillage, mechanized sowing, mechanized harvesting and mechanized irrigation. Considering that parts of the land are repeatedly cultivated in a year, a re-seeding coefficient (the ratio of sowing area to tillage area) is introduced to correct the workload statistic of the mechanized tillage when the data are input into the model. Table A2 presents the number and the total power of the agricultural machinery mentioned in this study.

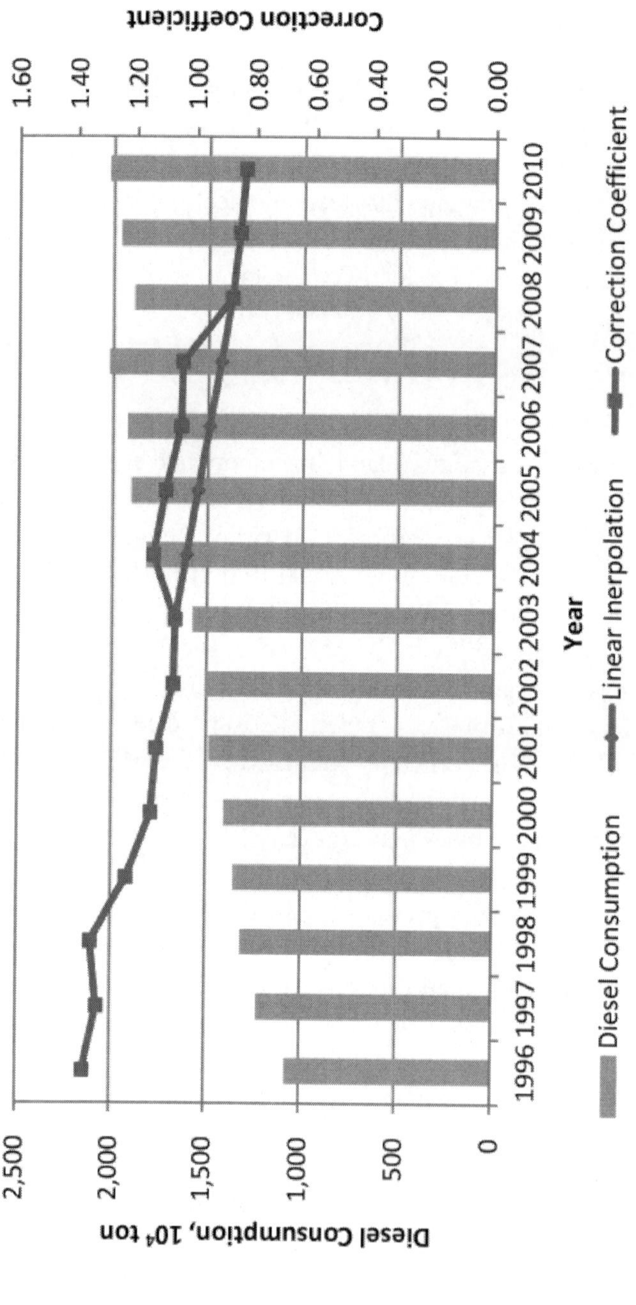

FIGURE 5: Correction coefficient and diesel consumption statistics.

TABLE 2: Unit diesel consumption (g/kWh) of the agricultural machinery.

	Machinery								
Year	Tractors	Harvesters	Irrigation	Primary Processing	Animal Husbandry	Fishery	Forestry	Transportation	Farmland Construction
1996	407.96	423.67	450.99	411.91	426.16	383.17	414.97	408.68	338.56
1997	405.21	419.81	446.92	408.40	422.47	374.80	412.29	404.72	344.16
1998	406.88	420.81	449.53	410.07	424.14	373.89	414.53	405.75	348.48
1999	397.07	409.81	439.71	400.63	414.33	362.89	405.37	395.98	339.26
2000	389.93	402.37	432.35	393.59	406.36	358.04	397.80	388.52	330.26
2001	388.29	399.76	430.27	392.30	405.35	357.19	398.05	386.61	325.88
2002	383.16	394.04	425.47	386.99	400.21	353.13	391.89	381.17	318.80
2003	382.76	391.78	425.62	386.70	399.85	350.62	394.08	380.37	315.59
2004	379.09	386.31	421.85	383.29	396.51	349.70	386.81	376.77	311.79
2005	375.32	380.44	418.10	379.67	392.81	342.10	385.62	372.60	307.53
2006	371.08	373.16	414.74	376.20	389.43	340.16	382.12	369.09	318.42
2007	367.09	367.66	411.02	372.58	385.57	336.61	378.09	365.04	312.34
2008	362.76	360.42	406.74	368.98	380.35	332.78	373.67	359.09	306.54
2009	359.45	355.74	403.54	366.50	377.30	330.13	372.37	355.66	301.60
2010	356.93	352.70	401.25	364.70	375.13	328.06	370.71	352.89	297.35

7.4 RESULTS AND DISCUSSION

7.4.1 CHANGES OF CORRECTION COEFFICIENT

Figure 5 presents the changes of the correction coefficient. The correction coefficient has declined from 1.37 in 1996 to 0.84 in 2010, with an average annual decline of 3.46%. The decline in the correction coefficient proves that technological progress has been affecting the unit diesel consumption and the unit productivity of the agricultural machinery positively. One also can find that the load factor has not been growing as expected.

The curve of the correction coefficient shows a significant linear downward trend. However, there are step changes in the correction coefficient corresponding to the step changes in the diesel consumption from 2004 to 2007. The step changes of the energy consumption statistics are widespread in most sectors during the Eleventh Five-Year period including the diesel consumption in agriculture. According to careful analysis, it is believed that the step changes cannot reflect the real energy consumption, and they probably result from either changes in statistical methodology or artificially adjusted energy consumption numbers. The latter is more likely the main reason due to the existence of the compulsory target for reducing energy consumption stated in the Eleventh Five-Year Plan of China government, so the correction coefficients from 2004 to 2007 are corrected using linear interpolation.

7.4.2 CHANGES OF UNIT CONSUMPTION

Table 2 provides the unit diesel consumption of agricultural machinery. The unit diesel consumption generally maintained a steady downward trend from 1996 to 2010.

- For tractors, the unit diesel consumption declined from 407.96 g/kWh in 1996 to 356.93 g/kWh in 2010, an average annual decline of 0.95%.
- For harvesters, it declined from 423.67 g/kWh to 352.70 g/kWh, an annual decline of 1.30%.

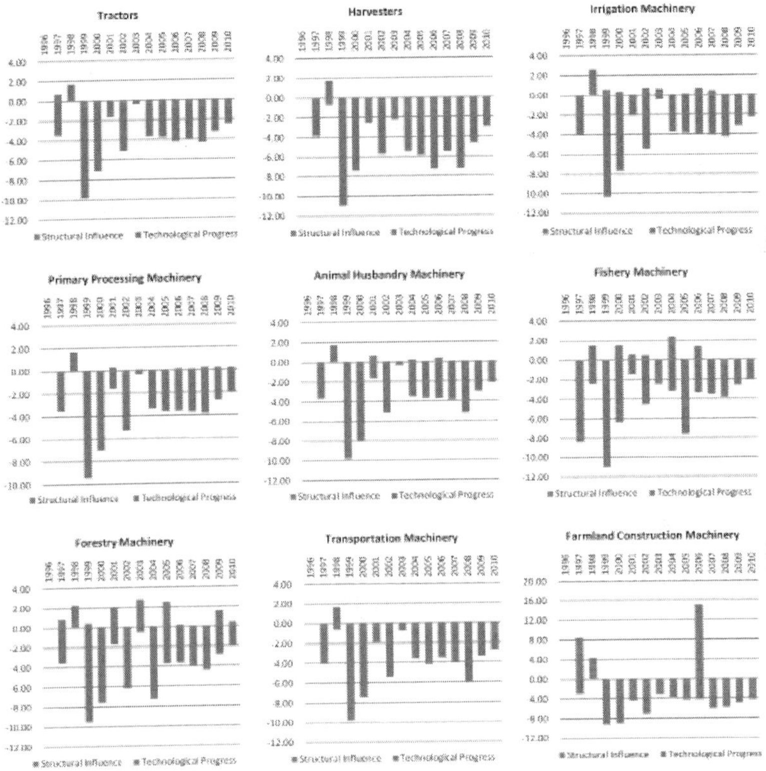

FIGURE 6: Changes in unit diesel consumption compared with that in the previous year.

- For irrigation machinery, it declined from 450.99 g/kWh to 401.25 g/kWh, an annual decline of 0.83%.
- For primary processing machinery, it declined from 411.91 g/kWh to 364.70 g/kWh, an annual decline of 0.87%.
- For animal husbandry machinery, it declined from 426.16 g/kWh to 375.13 g/kWh, an annual decline of 0.91%.
- For fishery machinery, it declined from 383.17 g/kWh to 328.06 g/kWh, an annual decline of 1.10%.
- For forestry machinery, it declined from 414.97 g/kWh to 370.71 g/kWh, an annual decline of 0.80%.
- For transportation machinery, it declined from 408.68 g/kWh to 352.89 g/kWh, an annual decline of 1.04%.
- For farmland construction machinery, it declined from 338.56 g/kWh to 297.35 g/kWh, an annual decline of 0.92%.

Figure 6 shows the changes in the unit diesel consumption compared with that in the previous year. The red part shows the changes in the unit diesel consumption caused by technological progress. The blue part displays the changes in the unit diesel consumption caused by changes in machinery quantity at different intervals (structural influence). It is obvious that the effect of the technological progress plays a major role in almost all agricultural machinery. It can be predicted that the trend will not change in the foreseeable future. However, the effect of technological progress is no longer significant for some machinery with low unit diesel consumption, such as farmland construction machinery.

Another way to reduce unit diesel consumption is to adjust the structure of the distribution of agricultural machinery. Results show that most of the agricultural machinery, such as tractors, harvesters, animal husbandry machinery, fishery machinery, transportation machinery and farmland construction machinery are becoming larger in size and lower in unit diesel consumption with the development of the agricultural economy. This is not the case for irrigation machinery, primary processing machinery and forestry machinery.

7.4.3 CHANGES OF UNIT PRODUCTIVITY

Table 3 shows the unit productivity of the four mechanized operations. The unit productivity generally maintained a steady upward trend from 1996 to 2010:

- The unit productivity of the mechanized tillage increased from 0.08 ha/h in 1996 to 0.10 ha/h in 2010, an average annual increase of 1.86%.
- The unit productivity of the mechanized sowing increased from 0.25 ha/h to 0.30 ha/h, an average annual increase of 1.27%.
- The unit productivity of the mechanized harvesting increased from 0.06 ha/h to 0.19 ha/h, an average annual increase of 8.13%, which is the largest growth rate in these four operations.
- The unit productivity of the mechanized irrigation increased from 74.41 m3/h to 79.87 m3/h, an average annual increase of 0.51%.

TABLE 3: Unit productivity of the four operations.

Year	Mechanized Tillage (ha/h)	Mechanized Sowing (ha/h)	Mechanized Harvesting (ha/h)	Mechanized Irrigation (m^3/h)
1996	0.08	0.25	0.06	74.41
1997	0.08	0.25	0.06	75.35
1998	0.08	0.25	0.07	74.01
1999	0.08	0.26	0.07	75.03
2000	0.08	0.26	0.07	75.95
2001	0.08	0.26	0.08	76.74
2002	0.08	0.27	0.08	76.77
2003	0.08	0.27	0.09	76.05
2004	0.09	0.27	0.10	76.75
2005	0.09	0.28	0.12	77.33
2006	0.09	0.28	0.13	77.17
2007	0.09	0.29	0.15	77.40
2008	0.09	0.29	0.17	78.30
2009	0.10	0.30	0.18	79.28
2010	0.10	0.30	0.19	79.87

Figure 7 shows the changes in unit productivity compared with that in the previous year. The red part indicates changes in unit productivity caused by technological progress. The blue part indicates changes in unit productivity caused by changes in machinery quantity at different intervals (structural influence).

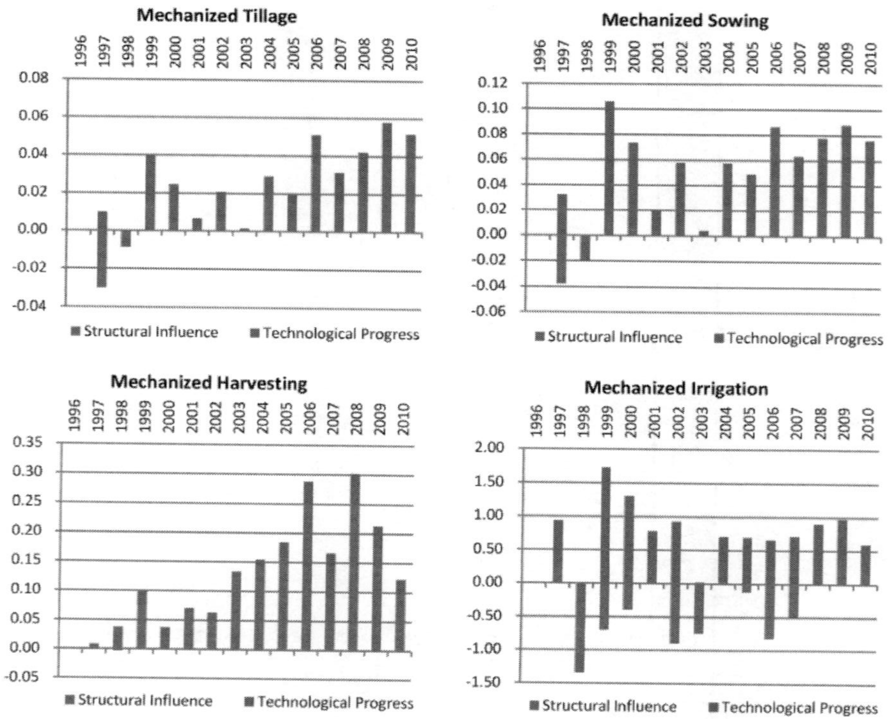

FIGURE 7: Changes in the unit productivity compared with that in the previous year.

Basically, technological progress has a positive influence on changes in the unit productivity of four operations. Improvement of unit productivity in mechanized harvesting is less than that in the other operations.

Structural adjustments to the machinery number caused an increase in the unit productivity of mechanized tillage, mechanized sowing and mechanized harvesting. Moreover, it also enables the unit productivity of the mechanized irrigation to decrease gradually with average rated power.

7.4.4 CHANGES OF WORKING HOURS

Table 4 shows the average annual working hours of agricultural machinery. Almost all average annual working hours of the agricultural machinery show a decreasing trend from 1996 to 2010, except for harvesters.

- The average annual working hours of the tractors declined from 56.33 h in 1996 to 41.00 h in 2010, an average annual decline of 2.24%.
- The average annual working hours of the harvesters increased from 87.19 h to 105.65 h, an average annual increase of 1.38%.
- The average annual working hours of the irrigation machinery declined from 217.24 h to 109.06 h, an average annual decline of 4.80%.
- The average annual working hours of the other machinery declined from 165.70 h to 114.43 h, an average annual decline of 2.74%.

Figure 8 shows the changes in annual average working hours compared with that in the previous year. The changes in average annual working hours could be caused by changes in actual workloads and changes in the unit productivity. In Figure 8, they are indicated by the red part and the blue part, respectively.

There is no doubt that an increase in unit productivity leads to a reduction in working hours. This is the case for all kinds of agricultural machinery. However, its effect is negligible compared with the effect caused by changes of actual workloads.

TABLE 4: Average annual working hours (h) of agricultural machinery.

Year	Tractors	Harvesters	Irrigation Machinery	Other Machinery
1996	56.33	87.19	217.24	165.70
1997	54.31	85.94	227.71	161.50
1998	55.07	91.48	195.59	163.48
1999	51.88	92.38	176.55	152.45
2000	49.58	89.13	158.42	144.64
2001	49.07	92.53	153.42	142.90
2002	47.49	91.69	143.85	137.53
2003	47.42	98.06	134.68	137.14
2004	46.29	101.17	134.90	133.55
2005	45.40	103.76	129.25	129.93
2006	44.45	107.22	127.96	126.28
2007	43.33	106.57	120.87	122.59
2008	42.38	107.16	115.53	118.86
2009	41.60	106.56	113.22	116.30
2010	41.00	105.65	109.06	114.43

Due to the effect of actual workloads, working hours of the tractors, irrigation machinery and other machinery continue to decline from 1996 to 2010. The harvesting machinery is the only one whose working hours have increased. This means that the total power of the agricultural machinery (excluding harvesters) grew faster than the actual workloads. Subsidies policy for purchasing agricultural machines upon 2004 results in the massive growth of the agricultural machinery at the expense of the waste of partial production capacity. Furthermore, economic life of most agricultural machinery does not exceed 15 years in China and a large number of scrapped agricultural machinery need to be recycle every year. Chinese energy policy maker should pay attention to this problem. Some materials such as aluminum and steel, are easily recyclable and thus their post-consumer recycling takes much less energy than production of finished materials from virgin feedstocks [24,25].

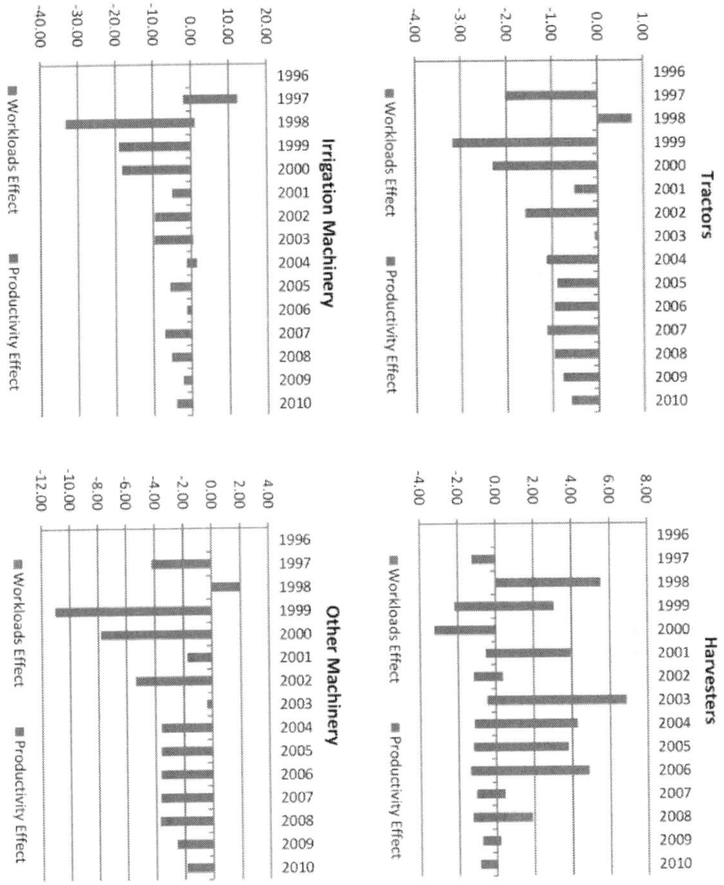

FIGURE 8: Changes in the annual average working hours compared with that in the previous year.

In addition, the average annual working hours of agricultural machinery powered by electricity is about 500–600 h in China. It is significantly higher than the working hours of machinery powered by diesel. It is believed that instability of the diesel supply and rising prices are the reasons for the low utilization rate of diesel machinery.

7.4.5 DIESEL FLOWS IN AGRICULTURE

Table 5 shows diesel consumption in different sectors of agriculture in China from 1996 to 2010.

- Diesel consumption in four farming operations increased from 652.31×10^4 ton in 1996 to 884.01×10^4 ton in 2010, with an annual increase of 2.19%. In these four operations, the mechanized harvesting has the largest annual growth rate (i.e., 19.39%) of the diesel consumption. The annual growth rates of mechanized tillage and mechanized sowing are 3.68% and 4.28%, respectively, which are slightly higher than the average level among all farming operations. Slight negative growth in the diesel consumption of mechanized irrigation was witnessed in the past thirteen years.
- Diesel consumption by primary processing increased from 107.88×10^4 ton in 1996 to 111.78×10^4 ton in 2010, with an annual increase of 0.25%.
- Diesel consumption by animal husbandry increased from 8.07×10^4 ton in 1996 to 17.87×10^4 ton in 2010, with an annual increase of 5.84%.
- Diesel consumption by fishery increased from 54.86×10^4 ton in 1996 to 68.35×10^4 ton in 2010, with an annual increase of 1.58%.
- Diesel consumption by forestry increased from 0.31×10^4 ton in 1996 to 3.95×10^4 ton in 2010, with an annual increase of 20.07%.
- Diesel consumption by transportation increased from 238.74×10^4 ton in 1996 to 860.53×10^4 ton in 2010, with an annual increase of 9.59%.
- Diesel consumption by farmland construction increased from 13.94×10^4 ton in 1996 to 76.61×10^4 ton in 2010, with an annual increase of 12.94%.

Figure 9 shows the diesel flows of the agriculture in China. It is obvious that farming and transportation are the two largest diesel consumers, while other sectors account for a negligible share. From 1996 to 2010, diesel consumption by farming grew smoothly while diesel consumption by transportation grew sharply. Transportation consumed nearly half of the total diesel in 2010 and its share can be expected to continue to grow in the future.

TABLE 5: Diesel consumption (104 ton) in different sectors of agriculture in China.

Year	Farming					Primary Processing	Animal Husbandry	Fishery	Forestry	Transportation	Farmland construction
	Total	Mechanized Tillage	Mechanized Sowing	Mechanized Harvesting	Mechanized Irrigation						
1996	652.31	214.53	33.76	13.61	390.40	107.88	8.07	54.86	0.31	238.74	13.94
1997	713.39	225.42	34.92	16.22	436.83	110.28	8.14	65.40	0.23	317.59	14.37
1998	699.85	245.02	37.87	21.38	395.58	117.83	9.49	71.79	0.21	399.54	16.00
1999	692.02	245.79	38.22	24.94	383.07	118.24	10.40	69.33	0.21	447.28	16.82
2000	672.33	248.06	38.63	27.25	358.39	120.66	11.23	69.32	0.71	508.50	22.26
2001	688.59	249.97	38.95	31.32	368.36	124.17	11.26	70.42	0.74	561.88	28.25
2002	668.75	248.18	38.76	34.91	346.90	123.14	12.09	67.08	0.98	602.70	32.76
2003	663.40	255.74	39.95	47.13	320.59	126.51	12.74	68.56	1.24	663.72	38.43
2004	693.71	266.36	41.98	55.06	330.32	120.78	12.89	64.65	1.59	705.20	42.67
2005	716.95	278.87	44.12	67.89	326.07	123.50	13.90	61.17	1.71	750.62	49.00
2006	747.16	290.89	46.86	83.09	326.32	121.20	13.70	64.36	1.91	743.74	57.19
2007	752.66	298.75	48.51	93.26	312.14	119.74	13.98	66.24	2.28	758.98	62.37
2008	803.33	327.90	53.90	113.20	308.33	116.04	17.27	66.43	2.66	810.93	71.23
2009	847.41	341.90	57.33	137.69	310.49	113.58	17.53	67.14	3.22	835.99	75.03
2010	884.01	355.64	60.69	162.68	304.99	111.78	17.87	68.35	3.95	860.53	76.61

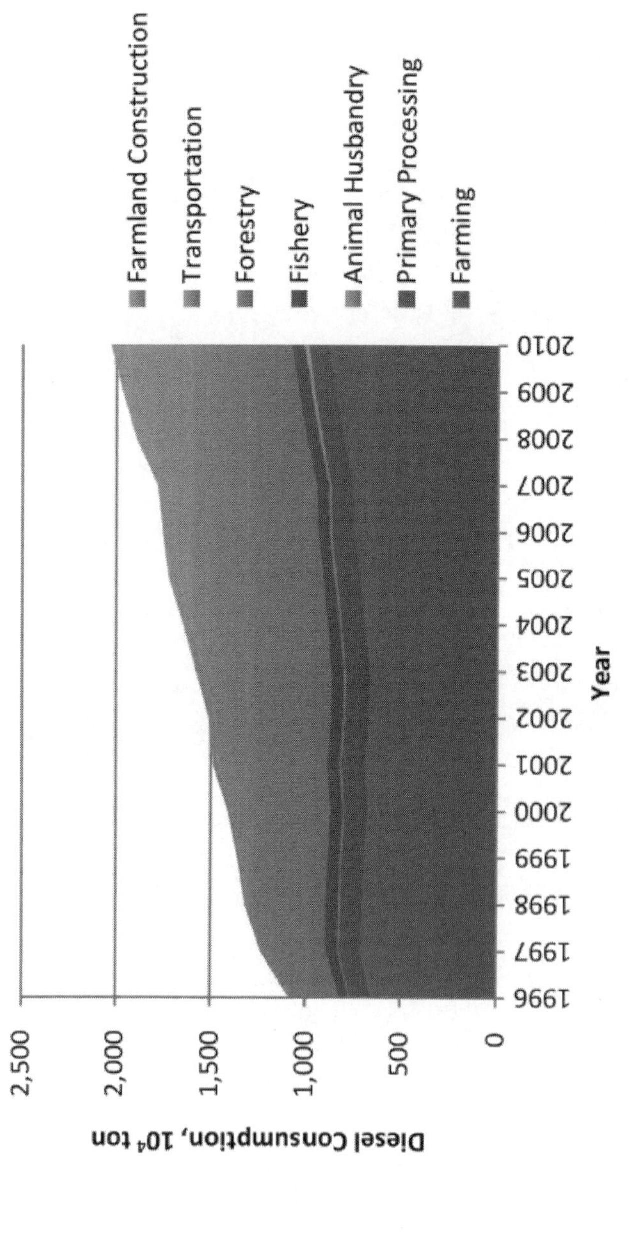

FIGURE 9: Diesel flows of the agriculture in China.

TABLE 6: Diesel consumption intensity (kg/ha) of the operations.

Year	Mechanized Tillage	Mechanized Sowing	Mechanized Harvesting	Mechanized Irrigation
1996	33.47	10.46	7.49	598.99
1997	34.16	10.19	7.90	607.62
1998	34.12	9.90	9.16	586.66
1999	32.66	9.51	9.75	567.48
2000	32.53	9.61	10.22	533.29
2001	32.84	9.49	11.66	552.11
2002	32.89	9.34	12.77	517.31
2003	33.89	9.79	17.19	476.54
2004	33.35	9.47	18.05	496.21
2005	33.00	9.36	19.73	483.05
2006	33.74	9.38	21.70	476.45
2007	33.11	9.20	22.11	443.55
2008	33.06	9.16	23.89	420.89
2009	32.72	8.82	25.83	416.73
2010	31.90	8.80	27.27	400.19

7.4.6 DIESEL CONSUMPTION INTENSITY

Table 6 shows diesel consumption intensity in different operations from 1996 to 2010. Little change occurred to the diesel consumption intensity of mechanized tillage and its value remained at around 33 kg/ha. The diesel consumption intensity of mechanized sowing declined slightly from 10.46 kg/ha in 1996 to 8.80 kg/ha. By an annual growth rate of 9.99%, the diesel consumption intensity of mechanized harvesting increased rapidly from 7.49 kg/ha in 1996 to 27.27 kg/ha in 2010. Because of the popularity of irrigation machinery powered by electricity, the diesel consumption intensity of mechanized irrigation which is far more than that of other operations declined from 598.99 kg/ha in 1996 to 400.19 kg/ha in 2010, with an annual decrease of 2.75%.

7.5 CONCLUSIONS

This study is a fundamental research for establishing the agricultural energy demand forecasting model. The simulation results demonstrate that the methodology used in this study is proper and accurate. The conclusions and the relevant policy recommendations are summarized as follows:

- For agriculture in China, farming and transportation are the two largest diesel consumers, accounting for 86.23% of the total diesel consumption in agriculture in 2010, while the other sectors account for a negligible share. Differing from the farming in this respect, more attention should be paid to the fast growth of the diesel consumption in the transportation in the forecasting model.
- Technological progress positively affected unit diesel consumption and the unit productivity of all machinery from 1996 to 2010. However, there is great potential in reducing unit diesel consumption and increasing unit productivity. The Chinese government should continue to promote technological progress and to improve in the field of mechanized agriculture.
- With the development of the agricultural economy, most of the agricultural machinery becomes larger and larger in size, more diesel fuel efficient and productive. However, irrigation machinery has proved to be an exception. Diesel consumption in mechanized agriculture can be reduced by preventing the miniaturization trend of irrigation machinery and raising the proportion of the medium-sized and large-scale agricultural machinery.
- The annual average working hours of the agricultural machinery (except harvesters) continue to decline from 1996 to 2010. Subsidies policy for purchasing agricultural machines upon 2004 leads to the massive growth of the agricultural machinery at the expense of the waste of the partial production capacity. This means that machinery sits idle in the yard for most of the time. Although this may not directly affect diesel fuel consumption, it directs resources to the manufacturing of agricultural machinery and increases the cost of the agricultural production.
- The annual average working hours of the agricultural machinery powered by diesel are about 40–120 h which is much fewer than that of agricultural machinery powered by electricity (i.e., 500–600 h). With an adequate power supply and feasible techniques, it is effective to save energy and improve utilization by replacing diesel machinery with electricity machinery.

REFERENCES

1. Dincer, I.; Hussain, M.; Al-Zaharnah, I. Energy and exergy utilization in agricultural sector of Saudi Arabia. Energy Policy 2005, 33, 1461–1467.

2. Utlu, Z.; Hepbasli, A. Assessment of the energy and exergy utilization efficiencies in the Turkish agricultural sector. Int. J. Energy Res. 2006, 30, 659–670.
3. Al-Ghandoor, A.; Jaber, J.O. Analysis of energy and exergy utilisation of Jordan's agricultural sector. Int. J. Exergy 2009, 6, 491–508.
4. Avara, A.; Karami, M. Energy and exergy efficiencies in agricultural and utility sectors of Iran compared with other countries. In Proceedings of the 2010 2nd International Conference on Mechanical and Electrical Technology (ICMET), Singapore, 10–12 September 2010; pp. 6–10.
5. Ahamed, J.U.; Saidur, R.; Masjuki, H.H.; Mekhilef, S.; Ali, M.B.; Furqon, M.H. An application of energy and exergy analysis in agricultural sector of Malaysia. Energy Policy 2011, 39, 7922–7929.
6. Rijal, K.; Bansal, N.; Grover, P. Energy in subsistence agriculture: a case study of Nepal. Int. J. Energy Res. 1991, 15, 109–122.
7. Ozkan, B.; Akcaoz, H.; Fert, C. Energy input-output analysis in Turkish agriculture. Renew. Energy 2004, 29, 39–51.
8. Hatirli, S.A.; Ozkan, B.; Fert, C. An econometric analysis of energy input–output in Turkish agriculture. Renew. Sustain. Energy Rev. 2005, 9, 608–623.
9. Alam, M.; Alam, M.; Islam, K. Energy flow in agriculture: Bangladesh. Am. J. Environ. Sci. 2005, 1, 213–220.
10. Uri, N.D. Energy and the use of conservation tillage in US agriculture. Energy Sources 1999, 21, 757–771.
11. Karkacier, O.; Gokalp Goktolga, Z.; Cicek, A. A regression analysis of the effect of energy use in agriculture. Energy Policy 2006, 34, 3796–3800.
12. Türkekul, B.; Unakıtan, G. A co-integration analysis of the price and income elasticities of energy demand in Turkish agriculture. Energy Policy 2011, 39, 2416–2423.
13. Daniëls, B.; Van Dril, A. Save production: A bottom-up energy model for Dutch industry and agriculture. Energy Economics 2007, 29, 847–867.
14. Baruah, D.C.; Bora, G.C. Energy demand forecast for mechanized agriculture in rural India. Energy Policy 2008, 36, 2628–2636.
15. Feng, L.J.; Shang, X.; Xing, Z. Survey analysis of wheel tractor fuel consumption [in Chinese]. Tractor Farm Transp. 2007, 34, 1–3.
16. GB/T15370. General Requirement of Agricultural Wheeled Tractors and Crawler Tractors; General Administration of Quality Supervision, Inspection and Quarantine of China, Standardization Administration of China: Beijing, China, 2004.
17. CSY China Statistical Yearbook; National Bureau of Statistics of China, National Development and Reform Commission: Beijing, China, 1997–2011.
18. CESY China Energy Statistical Yearbook; National Bureau of Statistics of China, National Development and Reform Commission: Beijing, China, 1997–2011.
19. CRSY China Rural Statistical Yearbook; National Bureau of Statistics of China: Beijing, China, 1997–2011.
20. CASR China Agriculture Statistical Report; Ministry of Agriculture of China: Beijing, China, 1996–2010.
21. CLRB China Land and Resources Bulletin; Ministry of Land and Resources of China: Beijing, China, 1996–2010.
22. CWRB China Water Resources Bulletin; Ministry of Water Resources of China: Beijing, China, 1996–2010.

23. Renpu, B. Statistical error analysis of operating level of Chinese agricultural mechanization Modern Agric. Equip. 2005, Z2, 72–76.
24. Cheremnykh, E.; Gori, F. Exergy and Extended Exergy Cost Assessment of a Commercial Truck. In Proceedings of the ASME 2010 International Mechanical Engineering Congress and Exposition (IMECE2010), Vancouver, Canada, 12–18 November 2010; Volume 5, pp. 461–468.
25. Seckin, C.; Sciubba, E.; Bayulken, A.R. An application of the extended exergy accounting method to the Turkish society, year 2006. Energy 2012, 40, 151–163.

There are several supplemental files that are not available in this version of the article. To view this additional information, please use the citation information cited on the first page of this chapter.

CHAPTER 8

ENERGY SAVINGS BY ADOPTING PRECISION AGRICULTURE IN RURAL USA

GANESH C. BORA, JOHN F. NOWATZKI, AND DAVID C. ROBERTS

8.1 BACKGROUND

Precision agriculture involves the development and adoption of knowledge-based technical management systems with the goal of optimizing application of fertilizer, chemicals, seeds, and irrigation resources to reduce input costs and maximize production. Precision agricultural technology has the ability to spatially vary the rates of all inputs to tailor to the varied production potential within the field, which can be gauged by use of geo-referenced historical crop yield data. Variable rate technology (VRT) can be used in both conventional and conservation tillage systems. Adoption of VRT technologies can reduce fuel use, since VRT coupled with GPS guidance systems reduces implement overlap during input applications, thus saving labor and machine hours [1].

The US agricultural energy use has increased tremendously in the last 50 years and accounts for approximately 17% of total national energy use.

This chapter was originally published under the Creative Commons Attribution License. Bora GC, Nowatzki JF, and Roberts DC. Energy Savings by Adopting Precision Agriculture in Rural USA. Energy, Sustainability and Society *2,1 (2012. DOI:10.1186/2192-0567-2-22.*

Approximately 400 gal of oil equivalent is used annually to produce the food that feeds each American; 19% of this is used to operate field machinery [2]. Precision technologies such as auto guidance reduce overlapping of passes while planting seeds or applying chemicals and fertilizers, which results in less fuel usage and labor time. The decreased application of chemicals and fertilizer also results in reduction of energy consumption by the agricultural machinery and reduced expenditure on inputs.

Natural Resource Conservation Service (NRCS) of the United States Department of Agriculture (USDA) estimates that even if only 10% of the US farmers use a guidance system for planting seeds in the USA, 16 million gallons of fuel, four million pounds of insecticide, and two million quarts of herbicide can be saved annually [1]. This will not only result in energy savings, but also financial savings for producers. A study by Clemson University showed that spatially varied tillage depth reduced energy requirements by 56% and fuel consumption by 34%. They also found that there is a potential energy savings of up to 52% by using variable rate irrigation systems [3].

Little research has been done relative to energy saving through precision agriculture, but USDA-NRCS has estimated the savings. A recent study by USDA notes that overlaps can be reduced from 24 to 2 in. by using a guidance system, which saves about US$13,000 in variable costs annually for a farm of 1,000 acres [1]. Based on these estimates, a GPS guidance system provides a substantial return on investment and pays for itself within one year. The return on investment increases with the size of the farm as the annual savings increase and the equipment cost is spread over more acreage. Provision of free GPS signals by the federal government has encouraged producers to use new precision tools, techniques, and services to enhance their efforts to save energy and reduce costs [1].

Different tillage systems can also play an important role in reducing fossil fuel use in farming operations. By practicing no-till farming, a farmer can save 3.9 gal/ac, or US13.65/ac, assuming diesel fuel costs of US3.50/gal [4]. Shibusawa [5] noted that the energy input–output ratio for crop production is very high, especially in fruits and vegetables. The machinery fuels and agro-chemicals derived from fossil fuels (including fertilizers, herbicides, and pesticides) are the major sources of high energy inputs in crop production systems.

This research has two major objectives: (1) to estimate the rate of adoption for GPS guidance and autosteering systems among agricultural producers of the Upper Midwest region of the USA and (2) to estimate the energy savings attributable to the adoption of these two precision agriculture technologies in crop production in the Upper Midwest region.

TABLE 1: Farm statistics of North Dakota in 2007

Items	Number
Number of farms	31,970
Land in farms	16,055,735 ha (39,674,586 ac)
Average size of farm	502 ha (1,241 ac)
Machinery and equipment value	US$174,683 per farm
Land and buildings value	US$957,053 per farm

(Adapted from NASS/USDA in 2007 [7])

8.2 METHODS

8.2.1 STUDY AREA

Upper Midwest region of the USA is comprised of rural states and the economy of the region mainly depends on agriculture. The state of North Dakota is a part of this region and situated in the latitude of 45°56′N to 49°00′N and longitude 96°33′W to 104°03′W. North Dakota experiences harsh, long winters, which usually begin in late November and continue through late March [6]. As a result, the crop cultivation period in North Dakota very short. But the farmers of the state have large capital resources with high land holding. Table 1 displays the farm statistics of North Dakota [7]. The average farm size is 502 ha (1,241 ac), and tremendous time and energy is spent in land preparation, planting, spraying, and harvesting using machines. Precision agricultural technologies like GPS guidance and autosteering systems help reduce overlapping of equipment and tractor passes to save fuel, labor, and time [1]. These technologies may be especially valuable in regions such as North Dakota, where fixed costs of equipment can be spread over large farms.

8.2.2 GPS GUIDANCE AND AUTOSTEERING SYSTEMS

GPS guidance and autosteering systems are used in agricultural equipment to increase operational efficiency and effective field capacity, resulting in more area covered per unit of time. The GPS guidance systems are used for parallel field operation with predetermined swath width across the field. It consists of GPS receiver, antenna, controller, and the display of choice, including either a light bar or a monitor. It can also have additional features such as a data logger, sound device, or visual display [8]. The GPS receiver can have differential corrections such as the Wide Area Augmentation Systems, the National GPS Differential Correction Service (beacon) or real-time kinematic differential correction, depending upon the required accuracy.

Autosteering systems in agricultural vehicles use the GPS guidance systems with the added option of automatically steering the vehicle. In this case, the mechanical device or an integrated electro-hydraulic control system installed in the cab automatically steers the vehicle based on the GPS signal and predetermined swath width. When using autosteer, the equipment operator only steers during turns and other maneuvers [9].

Though many farmers have been using these technologies for some time, having adopted them based on peer recommendations and anecdotal evidence of the reduced need for costly inputs, the actual benefits in terms of energy and time savings need to be quantified. This study estimated the savings of time and energy attributable to GPS guidance and autosteering systems in North Dakota to develop a precision technology program for the farmers.

8.2.3 RESEARCH METHODS AND DATA

Estimating energy use reduction attributable to precision farming technology is important for North Dakota farmers. A survey of about 1,000

Section I: Technology

1. Do you use **GPS Guidance systems** (including RTK) in your farming operations?
   Yes:____No: ____ (If no, go to question 2.)
   Estimate your annual savings from using GPS guidance for each operation: (Complete items that apply to your operation.)

Operation	Hours Saved	% Time Saved	Fuel Saved (Gallons)	Fuel Saved %
a. Tillage				
b. Planting				
c. Spraying				
d. Harvesting				
e. Other				

2. Do you use **autosteer** in your farming operations?
   Yes:____No: ____ (If no, go to question 3)
   Estimate your annual savings from using autosteer for each operation: (Complete items that apply to your operation.)

Operation	Hours Saved	% Time Saved	Fuel Saved (Gallons)	Fuel Saved %
a. Tillage				
b. Planting				
c. Spraying				
d. Harvesting				
e. Other				

FIGURE 1: Section 1 survey questions regarding technology adoption and perceived savings of time and energy.

Section II: Demography

Total acres operated:________________Total acres managed using precision agriculture:_________

1. Total household income:
   a. Less than $50,000 b. Between $50,000 and $100,000 c. More than $100,000
2. Number of farm workers (paid and unpaid including yourself): _________
3. County of Residence: ________________________

FIGURE 2: Section 2 survey questions regarding farm and farm household demographics.

farmers from different geographic regions of the state was undertaken with support from USDA-NASS, Fargo, North Dakota office. The farmers were selected at random by NASS. The target farmer demographic was the typical North Dakota crop producer, generally consisting of wheat, corn, and soybean growers. The survey consisted of a questionnaire with two main sections. In the first section, respondents were asked whether they had adopted specific precision agriculture technologies, including GPS guidance and autosteer, and how much savings of fuel and machine operator time they attributed to the use of these technologies throughout their operations. The second section of the questionnaire requests demographic data specific to the farm and farm household. Figures 1 and 2 exhibit the survey questions from sections 1 and 2 of the questionnaire, respectively.

The collected data were used to estimate the percentage of farm operators who have adopted GPS guidance systems and autosteer technology in their operations. We also used a logistic regression model [10] to determine how farm size and number of farm workers affect a farm operator's likelihood of using autosteer and/or GPS guidance. Lastly, we report the average producer's perception of the amounts of time and fuel saved by using these technologies in tillage, planting, spraying, harvesting, and other activities.

8.3 RESULTS AND DISCUSSION

Of the 1,000 questionnaires, only 60 were returned completed, which is a very low response rate—only 0.06%. This low response rate likely results from what is known as self-selection bias, which regularly causes difficulties in analyzing mail, internet, and telephone survey data [11]. In essence, self-selection bias means that respondents participate in surveys on topics that interest them. In the case of survey used in this research, we should expect that producers who currently use or are considering the use of GPS guidance and autosteering are more likely to choose to participate. Thirty-four percent of the respondents reported using GPS guidance systems in their agricultural vehicles, whereas 27% of the respondents reported using both GPS guidance and autosteering systems in their agricultural vehicles. Forty-eight percent reported using precision technologies in general to manage some amount of acreage. As a result of self-selection bias, these numbers likely overestimate the percentage of North Dakota farmers using these two precision technologies. In fact, the average farm size reported by respondents was 860 ha—much higher than the average size farm in North Dakota reported in Table 1—which indicates that operators of large farms were also more likely to respond. Table 2 presents the regression coefficients from a logistic regression model in which farm size and number of farms workers determine whether a producer uses any precision technologies. Note that the constant and the effect of farm size have statistically significant effects on the probability of using these technologies, while the number of farm workers has no significant impact. We also tested to see whether income category had significant explanatory power, which it did not. However, we dropped income from the model because the direction of causality (if it existed) between income levels and precision agriculture adoption is ambiguous. That is, income category may be endogenously determined. Based on the logistic regression model, the probability of a producer with an average size farm (502 ha for North Dakota) and two farm workers using Precision agriculture technologies is $P = 1/(\exp(-(-2.675 + 0.003 \times 502 - 0.027 \times 2)) + 1) = 0.227$. In other words, approximately 22.7% of farmers with average-sized holdings in North Dakota use precision agriculture technologies of some kind, including GPS guidance and/or autosteer. Assuming a symmetric, approximately normal distribution of farm size in North Dakota with a mean of 502 ha, 22.7% of farm operators have adopted some type of precision agriculture technology. Thus, the logistic regression model helps attenuate bias from respondent self-selection of the survey.

TABLE 2: Logistic regression results relating precision agriculture use to farm size and number of farm workers

Parameter	Description	Estimate	Standard error
α	Constant	−2.675[a]	0.787
β_1	Effect of farm size	0.003[a]	0.001
β_2	Effect of number of farm workers	−0.027	0.187

[a]*Statistical significance at the 99% confidence level or higher.*

Table 3 shows the amount and percentage of time and fuel savings farmers attributed to GPS guidance systems in their agricultural vehicles, while Table 4 shows savings they attributed to autosteering systems in North Dakota. The average respondent reported saving 65 h of machine operator time by using GPS guidance systems, which is about 6% of the farm operation time. GPS guidance systems also reduced fuel use by 1,647 l (435 gal), which is about 6.3% of fuel use for the average farmer in the sample. Based on these data and a conservative estimate of three dollars per gallon of fuel, the use of GPS guidance systems can save an average of US1,305 per farm in fuel costs alone.The use of autosteering systems can save an average of 75h of peak farming time, resulting in a 5.81,479 per farm in fuel costs. Based on only the estimated fuel savings, GPS guidance or autosteering systems can provide a positive return on investment within a year or two, depending on the brand name of the system used. However, producers using these systems also report saving time. If producers use hired labor, reduced machine operating hours can reduce the need for hired labor. Otherwise, the time savings allow the reallocation of family labor to other on-farm production activities, off-farm employment, and leisure, which are high-value activities. A conservative estimate for the value of an hour of labor saved is the hourly wage paid to displaced hired labor, which averaged US$11.29 per hour nationally in January 2011 [12]. Thus, the monetized values of time saved for the average farm are US733.85 and US851.27 for GPS guidance and autosteering systems, respectively.

TABLE 3: Average savings through GPS guidance systems

Operations	Time saved (hours)	Time saved (%)	Fuel saved (gallons)	Fuel saved (%)
Tillage	18.8	6.25	188	6.44
Planting	22.7	6.47	114	6.82
Spraying	17.5	7.11	86.6	8.00
Harvesting	6.00	4.33	46.7	4.00
Average total	65.0	6.04	435	6.32

TABLE 4: Average savings through autosteering systems

Operation	Time saved (hours)	Time saved (%)	Fuel saved (gallons)	Fuel saved (%)
Tillage	16.9	4.7	217	5.14
Planting	25.5	6.29	116	5.8
Spraying	23.0	6.33	59.9	5.38
Harvesting	10.0	5.67	100	5.00
Average total	75.4	5.75	493	5.33

8.4 CONCLUSIONS

The study was undertaken to estimate the farm energy savings in terms of fuel and time by using GPS guidance and/or autosteering systems in farm vehicles in the Upper-Midwest region of the USA. A survey was conducted in the state of North Dakota, and the results indicated that there is a good adoption of precision agricultural technology in the region. Thirty-four percent of the respondents used GPS guidance systems, resulting in savings of 6% of time and 6.32% of fuel. The results also showed that 27% of farms used autosteering systems and saved 5.75% of time and 5.33% fuel. Almost 1,870 l (500 gal) of fuel is saved using autosteering systems, and this is equivalent to savings of approximately US$1,500 per farm. Based on the perceptions of farmers who have adopted precision agriculture technology in the state of North Dakota, the two technologies investigated in this research provide a positive return on investment and would be beneficial to North Dakota's agricultural sector if adopted more widely.

REFERENCES

1. USDA-NRCS: Conservation practices that save: precision agriculture. 2006. Available at http://www.usada.gov. Accessed 14 Jul 2012
2. Pfeiffer DA: Eating fossil fuels. From The Wilderness Publications, Sherman Oaks; 2004.

3. Khalilian A: Precision agriculture research. Clemson Public Service. University of Clemson; 2009. Available via http://www.clemson.edu/public/rec/edisto/research/precision_ag.html. Accessed 13 Jul 2011
4. USDA-NRCS: Conservation Resource Brief: Energy Management. Number 0608. U. S. Department of Agriculture; 2006. Available: http://www.nrcs.usda.gov/feature/outlook/Energy.pdf. Accessed 26 Oct 2012
5. Shibusawa S (2003) Development of plant production system using mirror duct (Part I) - a strategy for conjunction of plant factory and precision agriculture in small scale farming. Paper no. 034072 presented at: ASAE Annual Meeting held at Las Vegas, Nevada, July 27–30, 2003. ASABE, St. Joseph, MI; 2003.
6. NDDES: North Dakota Department of Emergency Services: severe winter weather. 2012. Available at www.nd.gov/des/get/severe-winter-weather/. Accessed 26 Oct 2012
7. USDA-NASS: Census of agriculture: North Dakota State and County Data, 2007. 2009. http://www.agcensus.usda.gov/Publications/2007/Full_Report/Census_by_State/North_Dakota. Accessed 10 Oct 2012
8. Sullivan M, Ehsani MR: GPS guidance systems - an overview of the components and options. Ohio State University Extension Factsheet. AEX-570–02, Columbus, Ohio; 2002. Available at http://ohioline.osu.edu/aex-fact/0570.html. Accessed 13 Jul 2011
9. Adamchuk VI: Satellite-based auto-guidance. University of Nebraska at Lincoln Extension Circular, EC706, Lincoln, Nebraska; 2008. Available at http://ianrpubs.unl.edu/epublic/live/ec706/build/ec706.pdf. Accessed 13 Jul 2011
10. Greene WH: Econometric Analysis. 5th edition. Upper Saddle River, Prentice Hall; 2002.
11. Hudson D, Seah L, Hite D, Haab T: Telephone presurveys, self-selection, and non-response bias to mail and Internet surveys in economic research. Appl Econ Lett 2004, 11:237–240.
12. USDA-NASS: Farm labor report ISSN: 1949–0909. 2011. Available at http://http://usda01.library.cornell.edu/usda/nass/FarmLabo//2010s/2011/FarmLabo-02-17-2011.pdf. Accessed 26 Oct 2012

CHAPTER 9

REDUCING CARBON EMISSIONS THROUGH IMPROVED IRRIGATION MANAGEMENT: A CASE STUDY FROM PAKISTAN

ASAD SARWAR QURESHI

9.1 INTRODUCTION

Groundwater has emerged as an exceptionally important water resource, and growing demand for its use in agriculture, domestic and industrial contexts grades it as a resource of strategic importance. In view of the high evapotranspiration and salinity environment under which irrigated agriculture in the Indus basin is practised, the availability of surface water resources is only marginally sufficient for basin-wide, year-round high-intensity cropping (Bhutta and Smedema, 2007; Qureshi et al., 2009). This difference between crop water requirements and surface water supplies, combined with generally unreliable and relatively inefficient water distribution systems, has led to the exploitation of groundwater where conditions allow (World Bank, 2007; Qureshi et al., 2009).

Reprinted with permission from John Wiley & Sons. Qureshi AS. Reducing Carbon Emissions Through Improved Irrigation Management: A Case Study from Pakistan. Irrigation and Drainage ***63**,1 (2014); 132–138. DOI: 10.1002/ird.1795.*

The increasing role of groundwater in agriculture has made it very energy-intensive. Groundwater exploitation has enabled farmers to supplement their irrigation requirements and to cope with the vagaries of the surface supplies. This allows them not only to increase their production level and incomes but also enhance their opportunities to diversify their income base and to reduce their vulnerability to the seasonality of agricultural production, and to external shocks such as droughts (Bhutta, 2002; Qureshi et al., 2009). Groundwater use has also increased resilience to climate change because surface storages have fared poorly on these counts. These benefits will become even more important as climate change heightens hydrological variability. From society's point of view, aquifer storage is also advantageous because it minimizes water loss through non-beneficial evaporation for semi-arid countries like Pakistan, where surface storages can lose 3 m or more of their storage every year through pan evaporation (Shah, 2009).

The introduction of cheap technologies has played a key role in the groundwater boom in Pakistan. As a result, farmers tend to over-irrigate and a considerable amount of pumped water evaporates, or goes back to the aquifer through deep percolation. In both ways, a significant amount of consumed energy does not contribute to biomass production (Karimi et al., 2012). Other disadvantages of excessive groundwater use are declining groundwater tables and increasing salt content in the pumped groundwater. Groundwater irrigation is also expensive as compared to gravity-run canal irrigation. Furthermore, groundwater irrigation is also considered an environmental hazard because the energy used in pumping groundwater directly contributes to CO_2 discharge (Shah, 2009).

Pakistan is one of the lowest carbon emitters in the world but the increasing use of groundwater for irrigation is putting extra pressure on energy resources and directly contributes to an increase in CO_2 discharge. Therefore, productive and efficient use of groundwater at farms and decreasing pumping is beneficial for stabilizing aquifers and reducing carbon emissions, which could be a key climate change adaptation strategy. This paper estimates the CO_2 emissions as a result of groundwater extraction and quantifies reductions in energy consumption and CO_2 emissions through the adoption of improved irrigation management strategies.

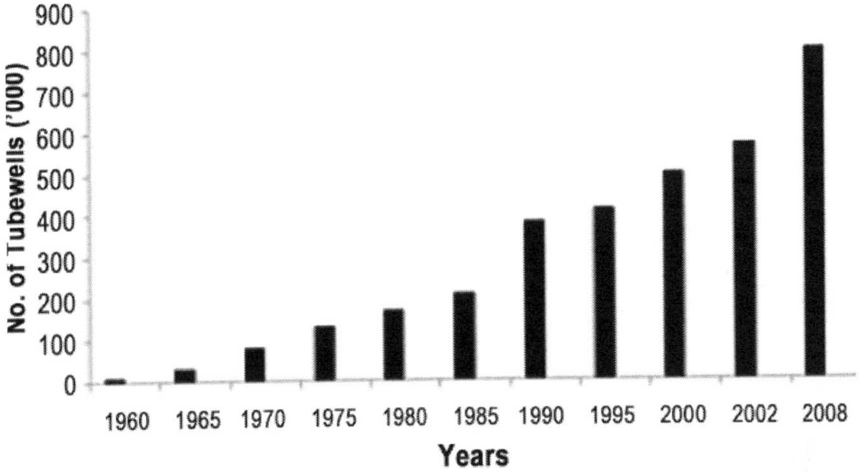

FIGURE 1: Development of private tubewells in the Punjab Province (Data source: Punjab Irrigation Department)

9.2 OVERVIEW OF GROUNDWATER IRRIGATION IN PAKISTAN

9.2.1 GROUNDWATER EVOLUTION IN PAKISTAN

The use of groundwater for irrigated agriculture in Pakistan has a long history. Before 1960s, groundwater extraction was carried out by means of open wells with rope and bucket, Persian wheels, karezes, reciprocating pumps and hand pumps. Large-scale extraction and use of groundwater for irrigated agriculture in the Indus basin started during the 1960s with the launching of Salinity Control and Reclamation Projects (SCARPs). Under this public sector programme, 16 700 wells (supplying an area of 2.6 million ha) with an average capacity of $80\,l\,s^{-1}$ were installed to control groundwater and salinity problems (Bhutta and Smedema, 2007).

The demonstration of SCARP tubewells was followed by an explosive development of private tubewells with an average discharge capacity of about $28\,l\,s^{-1}$. The provision of subsidized electricity by the government and the introduction of locally made diesel engines provided an impetus for a dramatic increase in the number of private tubewells. Currently, about 1.2 million small-capacity private tubewells are in operation in Pakistan (Qureshi et al., 2008). Out of these, 800 000 are located in Punjab (Figure 1). Investments in the installation of private tubewells are of the order of US$400 million whereas the annual benefits in the form of agricultural production are to the tune of US$2.5 billion (Shah et al., 2003). The estimated number of users is over 2.5 million farmers, who exploit groundwater directly or hire the services of tubewells from their neighbours. Groundwater currently provides more than 50% of the total crop water requirements, with flexibility of availability on an as and when needed basis (Shah, 2007).

9.2.2 PATTERNS AND BENEFITS OF GROUNDWATER USE

In Pakistan, about 70% of the private tubewells are located in the canal command areas where groundwater is used in combination with the canal water, whereas the rest provide irrigation based on groundwater alone. The combined use of surface water and groundwater (usually referred as conjunctive use) is now practised on more than 70% of the irrigated lands in Pakistan. The area irrigated by groundwater alone has increased from 2.7 to 3.4 million ha, whereas the area irrigated by canal water alone has decreased from 7.9 to 6.9 million ha (Qureshi et al., 2004). In Pakistan large-scale production of major crops such as wheat, cotton, rice and sugar cane is only possible because of the supplemental use of groundwater for irrigation. The average cost of irrigating with groundwater is 30 times higher than that of surface irrigation (World Bank, 2007). The cost of canal water per year per hectare is US$5.5, whereas groundwater is marketed as US$67 ha^{-1} yr^{-1}.

The benefits of groundwater in Pakistan are multidimensional and range from drinking water supplies for the urban and rural population, to economic development as a result of higher agricultural production. The role groundwater irrigation has attained in maintaining the agricultural boom is unique and vital and will expand further in future due to mounting pressure to grow more food and increasing incidences of drought in the region. Qureshi et al. (2003) have shown that more than 70% of the farmers in the Punjab depend directly or indirectly on groundwater to meet their crop demands. Therefore management of this resource requires high level of attention and commitment both from government agencies and from agricultural and domestic users.

9.2.3 SUSTAINABILITY OF GROUNDWATER RESOURCES

The unregulated and uncontrolled use of groundwater has diminished its relative accessibility. The trend of continuous decline of the groundwater table has been observed in many areas of the Indus basin, which illustrates the serious imbalance between abstraction and recharge. Figure 2 shows the changes in groundwater table depths over a period of 10 years (1993–2003) in the Punjab province. As a result, many wells have gone out of production, yet the water tables continue to decline and the quality deteriorates. Excessive exploitation of aquifers in fresh groundwater areas has resulted in falling water tables and groundwater has become inaccessible in 5 and 15% of the irrigated areas of Punjab and Balochistan provinces, respectively. Although no recent estimates exist, it was estimated that under the "business as usual scenario," this area is expected to increase to 15% in Punjab and 20% in Balochistan by 2020 (Punjab Private Sector Groundwater Development Project (PPSGDP), 2000). The variation between different canal commands is mainly linked to groundwater quality. In relatively fresh groundwater areas, extraction is greater because farmers there tend to grow water-intensive crops such as rice and sugar cane. In poor-quality groundwater areas, extraction is low in order to avoid secondary soil salinization.

9.2.4 ENERGY USE FOR GROUNDWATER EXTRACTION IN PAKISTAN

In Pakistan, the use of electricity for groundwater pumping started in the 1970s, when the rural electricity grid was expanded and the government provided much-needed incentives for farmers to install tubewells to boost agricultural production. In 1980s, the tubewell population surged from 37

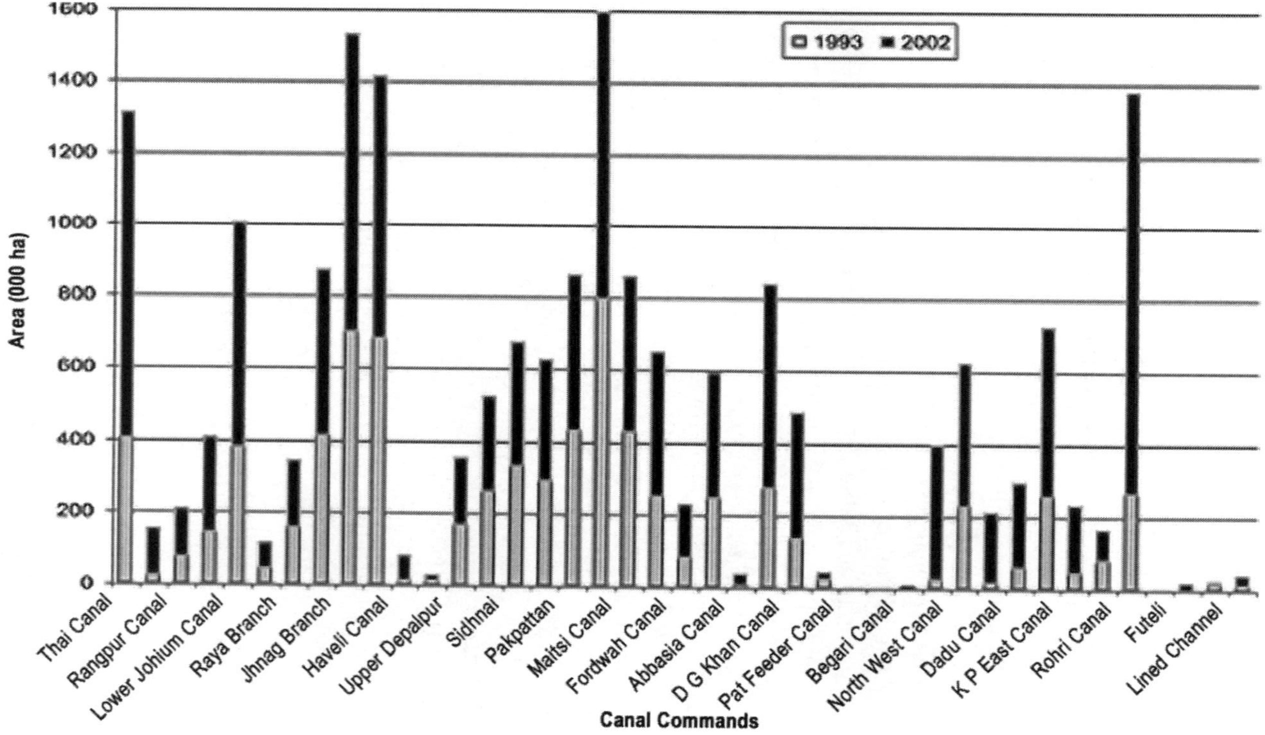

FIGURE 2: Increase in area with a groundwater table depth of 300 cm over a period of 10 years (1993–2003) in different canal commands of the Punjab and Sindh provinces (Source: Qureshi et al., 2009). This figure is available in colour online at wileyonlinelibrary.com/journal/ird

000 to 84 000, making it difficult for the government to collect revenue through the metering system (Qureshi and Akhtar, 2003). Increased electricity prices and unannounced power cuts resulted in the stagnation of electric tubewells and an increase in diesel tubewells. Although the cost of water from diesel tubewells (2.20 US¢ m^{-3}) was still higher than electric tubewells (0.70 US¢ m^{-3}), diesel tubewells were preferred due to low initial installation and operational costs.

The latest estimates suggest that in 2010, farmers extracted 50 billion cubic metres (BCM) of groundwater through 1.2 million diesel and electric tubewells (Qureshi et al., 2010). Of this, about 0.8 million are located in Punjab. About 200 000 tubewells are operated by electric motors whereas the remaining 1 million are run by diesel engines of various capacities. Out of a total 50 BCM of groundwater extraction, about 12 BCM is extracted using electric pumps and the remaining 38 BCM using diesel pumps.

The depth to groundwater is directly linked to energy requirements for water extraction. In a countrywide survey of 1200 private tubewells, Qureshi et al. (2003) found that in Pakistan, electric tubewells are used to extract water from greater depths (40–80 m) and diesel tubewells are used for shallow water table areas (6.0–15 m). The farmers use pumps which are not energy-efficient due to low capital investment. Due to high friction losses in wells and inefficient water conveyance systems, energy losses are very high. Energy requirements for extracting groundwater are highly sensitive to the dynamic head over which the groundwater is lifted. Therefore for energy calculations for this paper, a conservative estimate of dynamic head for electric and diesel pumps has been taken. For electric tubewells, a dynamic head of 60 m is assumed. For diesel pumps, a dynamic head of 10–15 m is considered because beyond this depth diesel pumps become extremely inefficient, forcing irrigators to switch to electricity. Therefore for diesel pumps, operational hours are more important for energy requirement calculations than dynamic head.

Electricity consumption in groundwater irrigation can be calculated based on the energy requirement to lift the water. To lift 1000 m^3 water from 1-m depth at 100% efficiency (without considering friction losses), 2.73 kWh of energy are required (Karimi et al., 2012). Thus energy consumption can be calculated as follows:

$$Ec = 2.73 \times D \times V/OPE \times (1\text{-}T1) \times 1000 \quad (1)$$

where:

Ec = electricity consumption (kWh)
D = lifting height (m)
V = volume (m^3)
OPE = overall pumping efficiency, and
Tl = transmission and distribution losses
(only in the case of electric pumps; otherwise zero).

The average overall pumping plant efficiency (OPE) of electric pumps in Pakistan is about 40% (Buksh et al., 2000). Electricity transmission and distribution losses are usually taken as 25% (Water and Power Development Authority (WAPDA), 2009). Therefore, electricity that is actually used to lift 1000 m3 of water from 1 m depth is 9.1 kWh. If we consider an average dynamic head of 60 m, then lifting 12 BCM of groundwater would require 6.0 billion kWh of electricity. This estimate is highly sensitive to the assumption about the dynamic head over which a representative electric pump lifts water.

Diesel-powered tubewells are even less efficient but they lift water to a smaller head; moreover, diesel does not face the transmission and distribution losses that electricity suffers and a litre of diesel provides the equivalent of 10 kWh of energy. Diesel tubewells are usually installed in shallow groundwater table areas (6.0–15 m). The fuel consumption of diesel engines (Chinese and slow speed diesel engines) is 1.5–2.5 l h^{-1} whereas tractor-operated tubewells burn 3.5–5.0 l h^{-1} (Qureshi et al., 2003). The utilization factor of private diesel tubewells is between 10 and 15% (1350 h yr^{-1}). Therefore total annual fuel consumption of 1 million diesel tubewells (assuming 2.5 l h^{-1} and 1350 h yr^{-1}) would be 3.5 billion litres. Therefore total energy consumption for groundwater extraction amounts to 41 billion kWh. Taking into account the consumption of 6 billion kWh electricity and 3.5 billion litres of diesel, it can be calculated that on average extracting 1 m^3 groundwater requires 0.820 kWh of energy in Pakistan. This amount of energy is equivalent to lighting up a 100 W bulb for more than 8 h.

9.2.5 CARBON FOOTPRINTS OF PAKISTAN'S GROUNDWATER IRRIGATION

Pakistan's contribution to total global greenhouse gas (GHG) emissions is miniscule (about 0.8%) and its per capita GHG emissions stand at a level which corresponds to one-third of the global average (Planning Commission, 2010). The total GHG emissions of Pakistan in 1994 were 182 MMT of CO_2 equivalence, which increased to 309 MMT of CO_2 equivalence in 2008, registering an increase of 3.9% yr^{-1} (Pakistan Atomic Energy Commission, 2009). The biggest contributor to GHG is the energy sector with 51% share, followed by the agriculture sector (39%), industrial processes (6%) and other activities (5%). Future estimates suggest that due to increasing energy demand, CO_2 emissions from the energy sector will increase to 2685 MMT of CO_2 equivalence from the current level of only 157 MMT of CO_2 equivalence. This shows the importance for Pakistan that it take serious steps to control GHG emissions in the energy sector. Controlling groundwater extraction could be one of the most effective strategies in this direction.

Carbon intensity of electricity and diesel is 0.4062 kg C kWh^{-1} and 0.732 kg C l^{-1}, respectively (Shah, 2009). This implies that annually a total sum of 3.8 MMT of CO_2 is emitted as a result of groundwater irrigation in Pakistan. Of this figure, which is roughly 1.2% of Pakistan's total carbon emissions, 1.4 MMT of CO_2 is emitted through electricity consumption and 2.4 MMT of CO_2 through diesel combustion. In other words, on average, the extraction of every cubic metre of groundwater in Pakistan comes with a hidden environmental cost of 80 g of carbon emissions. Therefore controlling energy demand in the agriculture sector would be a big step forward in limiting overall carbon emissions.

9.3 POTENTIAL FOR REDUCING CO_2 THROUGH IMPROVED IRRIGATION MANAGEMENT

There are different potential ways of reducing energy use in agriculture. The first option is to improve energy efficiency by increasing overall pumping plant efficiency through the use of high-quality pumps and electric motors. However, such interventions are expensive and, more importantly, have limited scope. The second option is to introduce on-site renewable energy sources such as wind and solar energy. These sources will neither lead to transmission and distribution losses, like electric energy, nor will they produce CO_2 emissions, like diesel tubewells. The initial investments in these resources might be high; however, considering their long-term economic and environmental benefits they should be given serious consideration. The third option is to reduce irrigation water demand through improved on-farm water management practices. This option is particularly relevant to Pakistan where on-farm water use efficiencies are extremely low. Average crop yields of major crops are low in Pakistan, for example: 2770 and 3190 kg ha^{-1} for wheat and rice, respectively. There is great variability in crop yields with some farmers achieving 5500 kg ha^{-1} of wheat and 3545 kg ha^{-1} of rice (Qureshi et al., 2004). The productivity of water in Pakistan is among the lowest in the world. For wheat, for example, it is 0.6 kg m^{-3} as compared to 1.0 kg m−3 in India. Maize yields in Pakistan (0.4 kg m^{-3}) are nine times lower than those in Argentina (2.7 kg m^{-3}) (Bastiaanssen, 2000). This reveals substantial potential for increasing water productivity.

9.3.1 IRRIGATION PRACTICES IN PAKISTAN AND OPTIONS FOR IMPROVEMENT

Despite the shortage of water, over-irrigation is a major problem in Pakistan. The impact of this is not only wastage of water, which could be used by other sectors or used in expansion of agriculture, but also waterlogging and soil salinity problems. This means that a significant amount of the applied irrigation water is lost by seepage from the irrigation canals and deep percolation in the fields (Bhutta and Smedema, 2007). Even though much of this lost water is now captured by extensive groundwater pumping and used downstream, this does not apply to the saline groundwater zone. From a basin perspective, improvements in farm irrigation efficiency may result in little gain in saving water except for those areas where groundwater is saline (Clemmens and Allen, 2005). Nevertheless, reducing water delivery to farms and improving farm water use efficiency are important from the perspective of other considerations like reducing energy consumption, costs and improving production (Karimi et al., 2012).

Farmers' current irrigation practices in Pakistan are aimed at applying the maximum amount of water in an attempt to maximize their crop yields. Farmers having access to groundwater in addition to canal water tend to apply more water compared to those who are fully dependent on canal water. Due to uncertainties in canal water supplies, farmers usually do not plan their irrigations in advance. Their decision to irrigate mainly depends upon the crop water need and availability of water in the canal system and/or access to groundwater. The water requirements of different crops depend upon environmental conditions, soil types and other factors that are equal across all the farms. However, different studies have shown that the number of irrigations applied to a wheat crop varies from 4 to 7, to cotton from 4 to 8, and to rice from 16 to 25 (Vlotman et al., 1994; Raza and Choudhry, 1998). The depth of individual irrigation applications has been the subject of many research studies. Vehmeyer (1992) found that it ranged from 60 to 90 mm. Vlotman and Latif (1993) determined the average depth applied per irrigation at between 70 and 80 mm. On the basis of field measurements, Raza and Choudhry (1998) reached a value of 60–90 mm with an average of about 85 mm per irrigation. If, on average, 6 irrigations to wheat and cotton and 20 irrigations to rice crop are considered with an amount of 80 mm per irrigation, irrigation water applied to wheat and cotton will be equal to 480 mm whereas for rice it will be 1600 mm. The average irrigation application in the Indus basin is 36% (Ahmad, 2009).

Considering the water scarcity in the Indus basin, many researchers have tried to find optimal irrigation schedules for different crops. The modelling work of Qureshi and Bastiaanssen (2001) has suggested that applying 300 mm of water to wheat and cotton (instead of the current practice of 420 mm) is enough to produce optimal crop yields without increasing salinity levels in the soil. This saving can be achieved by reducing amounts of individual irrigations. Based on their field experiments, Choudhary and Qureshi (1991) have also shown that improved irrigation management techniques such as furrow-bed and furrow-ridge can reduce irrigation requirements by 40%. They have recommended an irrigation application of 260–300 mm for wheat and cotton crops to achieve optimal yields.

TABLE 1: Comparison of total water use and water savings under current and improved irrigation practices (Data source: Punjab Agriculture Department)

Crop	Area (ha)	Current irrigation practices		Improved irrigation practices		Total water saving (BCM)
		Irrigation (mm)	Total water use (BCM)	Irrigation (mm)	Total water use (BCM)	
Wheat	8 578 000	480	41.2	300	25.7	15.5
Cotton	3 100 000	480	14.9	300	9.4	5.5
Rice	1 016 000	1 600	16.3	1 300	13.3	3.0
Total			72.4		48.4	24.0

Prathapar and Qureshi (1999) used the Soil–Water–Atmosphere–Plant (SWAP) model (Van Dam et al., 1997) to simulate optimal irrigation schedules for wheat and cotton crops. They found that irrigation applications can be reduced to 80% of the total crop evapotranspiration (ET) without compromising on yields and soil salinization, and recommended 300–320 mm as the optimal irrigation amount for wheat and cotton crops. Similarly improved irrigation methods for rice such as direct seeding also reduce irrigation amounts by 15–20% (Qureshi et al., 2006). The amount of water applied to rice was 1200 mm as compared to the 1870 mm usually applied under traditional planting. High efficiency irrigation methods such as drip and sprinkler systems have also proved successful in increasing water use efficiency. However, in a country like Pakistan where continuous availability of water and energy are big issues, adoption of these technologies will remain a challenge, especially for small farmers (Qureshi et al., 2010). For this reason, on-farm water conservation techniques which are less costly and energy intensive should be encouraged more.

To summarize the results of the above studies, these results suggest that 300 mm of irrigation water for wheat and cotton and 1300 mm for the rice crop is sufficient to produce optimal yields under the existing soil and climatic conditions of the Indus basin. Table 1 compares the irrigation amounts, total water use and water savings for current and optimized irrigation practices.

Table 1 clearly shows that adoption of the above-mentioned irrigation practices for wheat, cotton and rice can save up to 24 BCM of water, which is about 14% of the total renewable water available in the Indus basin. Applying these improved irrigation techniques to other crops can further reduce the water demand for irrigation and stress on groundwater. Under the current surface-water-scarce conditions of the Indus basin, this water is contributed through groundwater extraction, as is evident from the declining groundwater table conditions in most of the canal commands (Figure 2). Farmers with access to groundwater tend to apply more irrigation water than those farmers fully relying on surface water (Shah et al., 2003). Reducing groundwater extraction by 24 BCM will reduce diesel consumption by 2.2 billion litres (62%) and CO_2 emissions by about 40% (1.5 MMT of CO_2). With these reductions, total consumption of diesel will be reduced to 1.3 billion litres and CO_2 emissions to 2.3 MMT of CO_2. These calculations have been made assuming an irrigation application efficiency of 65%. Under the furrow irrigation method (the most widely practised in the Indus basin) irrigation efficiency ranged between 65 and 95% with an attainable level of 85% (United States Department of Agriculture (USDA)). Therefore greater water savings can be achieved by implementing optimized irrigation schedules together with advanced farm levelling, application rate control, and other management options.

The above analysis demonstrates that the adoption of improved irrigation practices will not only help in reducing energy consumption and CO_2 emissions but will be a big step forward in stabilizing aquifers. Adoption of these improved practices requires a shift in the thinking of farmers from "maximizing crop production" with increased irrigation supplies to "optimize crop production" with minimum irrigation supplies. Such a change in farmers' mentality could be facilitated by measures such as revising the existing energy pricing system. For instance, removing or limiting the subsidies on electricity could help to reduce groundwater over-pumping and encourage more efficient use of water.

9.4 CONCLUSIONS

Groundwater use in agriculture has increased significantly in the past few decades and it has become a lifeline to Pakistan's agricultural production. Currently, it provides more than 50% of the total water available at the farm gate and in many areas is the sole water resource for summer crops. However, rapidly dropping groundwater water tables in aquifers all over the country indicate that the extraction rate is far greater than the real capacity of these resources. Under these circumstances, groundwater availability might decrease considerably in future, which will have serious consequences for the food security of this country. On the other hand, groundwater use is also linked with a high energy demand and carbon footprint in Pakistan. In Pakistan, the extraction of 50 BCM of groundwater consumes 30 billion kWh of energy. Carbon emissions attributed to this energy use are 3.8 MMT of CO_2 yr^{-1}. Therefore reducing irrigation water demand through improved irrigation practices is vital for preserving the environment and sustaining groundwater resources.

Despite the fact that pumping is an energy-intensive activity, so far very little attention has been given to the carbon footprint of groundwater irrigation in Pakistan. This study shows that adoption of improved irrigation practices will save up to 24 BCM of irrigation water, which in turn, will reduce the energy demand and carbon emissions by 40%. This shows that enhancing water productivity through improved irrigation management can help in coping with water, energy, and climate change issues in Pakistan's agricultural sector.

ENDNOTES

1. Mostly privately owned diesel tubewells are powered by 10–24 hp engines. These engines are of two types, i.e. the 12–16 hp Chinese

engines known locally as 'Petter engines' and 20–24 hp slow speed engines known locally as the 'Black (Kala) engine'.

2. OPE is the product of power plant efficiency (engine, alternator, etc.), shaft efficiency and pump efficiency.

REFERENCES

1. Ahmad S. 2009. Water availability and future water requirements. Paper presented at the National Seminar on 'Water Conservation, Present Situation and Future Strategy. Ministry of Water and Power, Islamabad, Pakistan. May.
2. Bhutta MN. 2002. Sustainable management of groundwater in the Indus basin. Paper presented at the Second South Asia Water Forum, 14–16 December. Pakistan Water Partnership: Islamabad, Pakistan.
3. Bhutta MN, Smedema LK. 2007. One hundred years of waterlogging and salinity control in the Indus valley, Pakistan: a historical review. Irrigation and Drainage 56: 581–590.
4. Bastiaanssen WGM. 2000. Water issues for 2025: a research perspective. Research contribution to the World Water Vision report. Colombo, Sri Lanka: International Water Management Institute.
5. Buksh D, Ijaz H, Ahmad S, Yasin M. 2000. EMz ceramics for improving fuel efficiency of diesel pump sets in Pakistan. Journal of Nature Farming and Environment 1(2): 9–15.
6. Choudhary MR, Qureshi AS. 1991. Irrigation techniques to improve the application efficiency and crop yield. Journal of Drainage and Reclamation 3(1): 14–18.
7. Clemmens AJ, Allen RG. 2005. Impact of agricultural water conservation on water availability. In Proceedings of the EWRI World Water and Environmental Resources Congress, 2005: Impacts of Global Climate Change, 15–19 May, Anchorage, Alaska, USA; 14 pp.
8. Karimi P, Qureshi AS, Bahramlo R, Model D. 2012. Reducing carbon emissions through improved irrigation and groundwater management: a case study from Iran. Agricultural Water Management 108(2012): 52–60.
9. Pakistan Atomic Energy Commission. 2009. Greenhouse gas emission inventory of Pakistan for 2007–08. (Internal report – to be published.)
10. Punjab Private Sector Groundwater Development Project (PPSGDP). 2000. Legal and Regulatory Framework for Punjab Province. Technical Report No. 45. Lahore, Pakistan.
11. Prathapar SA, Qureshi AS. 1999. Modeling the effects of deficit irrigation on soil salinity, depth to watertable and transpiration in semi-arid zones with monsoon rains. International Journal of Water Resources Development 15(1/2): 141–159.
12. Qureshi AS, Bastiaanssen WG. 2001. Long-term effects of irrigation water conservation on crop production and environment in semi-arid zones. ASCE Irrigation and Drainage Engineering 127(6): 331–338.
13. Qureshi AS, Shah T, Akhtar M. 2003. The groundwater economy of Pakistan. IWMI Working Paper No. 64. International Water Management Institute, Colombo, Sri Lanka; 23 pp.

14. Qureshi AS, Akhtar M. 2003. Effect of electricity pricing policies on groundwater management in Pakistan. Pakistan Journal of Water Resources 7(2): 1–9.
15. Qureshi AS, Asghar MN, Ahmad S, Masih I. 2004. Sustaining crop production under saline groundwater conditions: a case study from Pakistan. Australian Journal of Agricultural Sciences 54(2): 421–431.
16. Qureshi AS, Masih I, Turral H. 2006. Comparing water productivities of transplanted and direct seeded rice for Pakistani Punjab. Journal of Applied Irrigation Science 41(1): 47–60.
17. Qureshi AS, McCornick PG, Qadir M, Aslam Z. 2008. Managing salinity and waterlogging in the Indus Basin of Pakistan. Agricultural Water Management 95: 1–10.
18. Qureshi AS, McCornick PG, Sarwar S, Sharma BR. 2009. Challenges and prospects for sustainable groundwater management in the Indus Basin, Pakistan. Water Resources Management 24(8): 1551–1569.
19. Planning Commission. 2010. Final Report of the Task Force on Climate Change. Government of Pakistan, Islamabad, Pakistan. 98 pp.
20. Raza ZI, Choudhry MR. 1998. Soil Salinity Trends under Farmer's Management in a Pipe Drainage Project. International Waterlogging and Salinity Research Institute (IWASRI) Publication No. 189. Lahore, Pakistan; 88 pp.
21. Shah T, Debroy A, Qureshi AS, Wang J. 2003. Sustaining Asia's groundwater boom: an overview of issues and evidence. Natural Resource Forum 27: 130–141.
22. Shah T. 2007. The groundwater economy of South-Asia: an assessment of size, significance and socio-ecological impacts. In Giordano M, Villholth KG (eds). The Agricultural Groundwater Revolution: Opportunities and Threat to Development. CABI Publications: United Kingdom; 7–36.
23. Shah T. 2009. Climate change and groundwater: opportunities for mitigation and adaptation. Environmental Research Letters 4(2009): 13.
24. United States Department of Agriculture (USDA). Irrigation efficiency. http://ddr.nal.usda.gov/bitstream/10113/4018/IND43939089.pdf.
25. Van Dam JC, Huygen J, Wesseling JG, Feddes RA, Kabat P, Van Walsum PEV. 1997. Simulation of Transport Processes in the Soil–Water–Air–Plant Environment. SWAP User's Manual. DLO-Winand Staring Centre: Wageningen, the Netherlands.
26. Vehmeyer P. 1992. Irrigation Management Strategies at the Farm Level in a Warabandi Schedule. IWASRI Publication No. 122. International Waterlogging and Salinity Research Institute: Lahore, Pakistan; 127 pp.
27. Vlotman WF, Latif M. 1993. Present and Suggested Water Management Strategies in Fixed Rotational Irrigation System at S1B9 of Fourth Drainage Project, Faisalabad. IWASRI Publication No. 148. International Waterlogging and Salinity Research Institute: Lahore, Pakistan; 90 pp.
28. Vlotman WF, Beg A, Raza ZI. 1994. Warabandi, theory and farmers practices at S1B9, Fourth Drainage Project, Faisalabad. Working paper, IWASRI Publication No. 141. Lahore, Pakistan; 281 pp.
29. Water and Power Development Authority (WAPDA). 2009. Hydropower Potential in Pakistan. Lahore, Pakistan. www.wapda.gov.pk
30. World Bank. 2007. Punjab Groundwater Policy—Mission Report. WB-SA-PK-Punjab GW mission report, June 2007. www.worldb¬–ank.org/gwmate.

PART IV

IMPACT OF POPULATION GROWTH

CHAPTER 10

IMPACT OF A GROWING POPULATION IN AGRICULTURAL RESOURCE MANAGEMENT: EXPLORING THE GLOBAL SITUATION WITH A MICRO-LEVEL EXAMPLE

A. H. M. ZEHADUL KARIM

10.1 INTRODUCTION AND CONTEXTUALIZING THE ISSUE

In the last few decades, there has been great concern over the issue of natural resource management in the global context. People are very much aware that the supply of various non-renewable natural resources on this planet is shrinking rapidly due to over-exhaustion and enhancement of resource appropriation. There has been rapid transformation of the world's natural landscape to agriculture, and it is learned that such use of natural resources will soon exceed its carrying capacity by causing an irreversible damage to its natural ecosystem. While land use practices often vary greatly across the world, their ultimate purpose usually remains the same which is to extract the natural resources for instant social needs, knowing clearly the severe impact of it on the environment. In the meantime, the world population has increased from 3 billion in 1959 to 6 billion in 1999, which took only 40 years for it to double. The US Census Bureau of the

This chapter was originally published under the Creative Commons Attribution License. Karim AHMZ. Impact of a Growing Population in Agricultural Resource Management: Exploring the Global Situation with a Micro-level Example. Asian Social Science *9,15 (2013). doi:10.5539/ass.v9n15p14. Table 4 was not part of the original article and was added with permission from the author.*

International Data Base also projected and mentioned that this number will be 9 billion in the year 2044, an increase of 50% within a span of 45 years (US Census Bureau, 2010). Accordingly, the demographers and environmentalists have posed a concern; the main challenge for the global environment is to determine our planet's capacity to sustain such a huge number of growing populations. In this context, the carrying capacity (Note 1) of the planet may further be measured by calculating the per capita requirement of food and nutrition subsistence. To provide adequate food subsistence to the people living with diverse diet will require at least 0.5 hectare of arable land per person (Lal & Steward, 1990), and at this time, we have only 0.27 hectare per capita land available to us, which will drastically be reduced to 0.14 hectare per person within the next 40 years due to loss of land caused by population pressure (Pimentel, 1993; Pimentel et al., 1994; Pimentel et al., 1995; Pimentel, 1997). In his book titled 'World Soil Erosion and Conservation' published in the year 1993, David Pimentel mentioned that per capita shortage in the availability of land has remained the major reason for severe food shortage and malnutrition in many parts of the world. The environmental depression is further intensified due to soil erosion in agricultural areas where 75 billion of metric tons of soil are demolished from the fields through wind and water, mostly affecting the cultivatable land (Myers, 1993). Furthermore, it is documented that deforestation and desertification have been occurring in the last two decades causing the human beings to be more vulnerable to shortages of land (Skole & Tucker, 1997). In the process of deforestation and desertification, more forest areas are converted for required farming activities (Note 2).

Based on the foregoing contextual introduction, the main purpose of this paper is to assess specifically the impact of growing population on available agricultural resources around the world which creates pressure on indigenous and sustainable agricultural management. Purposively therefore, the paper has four-fold collectives to relate demography with the available natural resources. At the initial point, the paper provides an outline of quantitative documentation showing a linear increase of the world's population in recent times. While the population grows at an unexpectedly speedy geometrical configuration as theoretized by Thomas Robert Malthus (1798), it is quite certain that this huge number of population will require food, water, and settlement which will force the world to place unexpected pressure on natural resources and land (Note 3). A comprehensive research on the world's major land use practices in recent times indicates that at least one-third of the world's land surface is now used for agriculture and millions of acres of land every year are converted to cultivation (Foley, 2005).

From this perspective, the paper will briefly provide information in regard to loss of natural resources annually for farming purposes. To exemplify such a trend in agricultural land use, the paper in its later section will show the direct impact of population growth on a farming community. More indicatively, the author depicts clearly the transitional transformation of the indigenous technique of farming to modern cultivation, requiring heavy demand on irrigation and chemical fertilizer which subsequently degrade the original fertile land and environment at the local level; such is the demographic impact on agricultural resource management. The paper finally concludes with a very modest caution for all inhabitants of this planet, saying that a judicious use of the environment is fully dependent on the honest formulation of accurate policies to keep population below replacement level. This is very much an essential and strategic requirement in our indigenous way of survival and resource management.

From the methodological point of view, the paper has adopted a conjunctive technique of triangulation where the secondary sources of data from global perspective have conceptualized an analytic-descriptive framework for explaining the world's demographic situation, contextualizing its impact on environmental resource management. In consonance with the above dimension, the paper has incorporated an ethnographic documentation at the micro-level, showing the situation at the village level. As a matter of fact, this is an important test on the effectiveness and accuracy of what has been stated at the macro perspective in the global context and an ethnographic brief in this context is proving our statement at the field level.

10.2 POPULATION GROWTH AROUND THE WORLD: ITS EQUATION AND IMPACT ASSESSMENT

Since the beginning of human history through the early 1800s, the global population had been increasing more or less at a consistent rate and as such, it did not pose any serious threat for the people around the world until that time. The overall statistics on world population exhibit that it remained at 1 billion until 1830, and it took 100 years to double the population to 2 billion in the year 1930. But subsequently, within a range of 30 years later in 1960, the cumulative growth of population stood at 3 billion worldwide. It took only another 15 years in 1975 to increase the total population of the world to 4 billion. To explain more analytically, the causative factor for a lower number of population growth prior to 1930 was not for its lower birth, but because of the high rate of mortality due to some severe and uncontrollable epidemics which caused a huge number of world population to perish at that time. The mortality rate in those days further increased due to conditions like famines, accidents, etc., which reduced human population even though there was high fertility. It is clear from the above statistics that within a stipulated period of only 45 years from 1930 to 1975 the population of the world had simply doubled. In this continuous process, twelve years later in 1987, the population of the world reached 5 billion. In the year 1999, it became 6 billion which further increased to 6.8 billion in the year 2009 (Population Reference Bureau, 2009). World population is expected to grow to 8.9 billion in 2050, and much of the demographic change up to this period will occur in the less developed nations. Although at present the population growth rate is 76 million a year, it does not seem to be appreciating much.

Although the overall population growth rate throughout the world had decreased considerably during the later part of the last century, the population growth rate in general remained consistently high in many poor and underdeveloped countries. During 1960, the population growth rate throughout the world peaked at 2.4% per year which was later reduced to 1.8% during 1999 meaning there is an increase of 87 million people every year. Extrapolating on such trend of lower fertility rate, UNDP's projected data clearly indicate that the world population will reach 9.1 billion by 2050 (Wright, 2008; UNDP, 2009). Rising population has already been a problem for many poor and developing nations of the world. For instance, the population in India which is approximately 1.8 billion inhabits an area of 3,287,240 sq. km. Its current growth rate is 1.9% per year which will double in the next 37 years (PRB, 1995; as quoted in Pimentel, 1998). Similarly, China has 1.27 billion people with a growth rate of 1.1% which is the optimum desirable rate of the percent size (Qu & Li, 1992). Despite the government's effort to reduce the growth rate by allowing only one child per couple, the population of China has continued to increase every year. One of the poorest countries of the world, Bangladesh has about 153 million people living in a surface area of 147570 sq. km. In 1930, its total population was only 35.5 million which has now increased to more than four times to make it 153.50 million in 2008; Bangladesh now faces a daunting challenge to feed its population where at least half of them are living in food based poverty level (Cuffaro, 1997; Karim, 2011).

TABLE 1: Global population situation

Year	Population (in Billion)
1830	1
1901	1.4
1930	2
1960	3
1975	4
1987	5
1999	6
	Total Fertility Rate
1969	6
1999	3
	Population Growth Rate
1969	2.4
1999	1.9
World population projected for 2050: 9.1 billion	

Sources: Prema Ramachardaran et al.; 2008; UNDP 2009. Rearranged and Modified

The total population of China, India, Bangladesh, Pakistan, Nigeria, Indonesia, Brazil and Ethiopia account for almost half of the total world population (UNDP, 2008). A few Asian (e.g. Malaysia) and Middle Eastern countries (e.g. Iraq, Iran, Saudi Arabia & Turkey) do not have any demographic problem in regard to their population size until now, as they possess sufficient land and natural resources which provide them with enormous economic prosperity. Yet it is suspected that they might face problems due to an increase in population in the near future. As demographers, we may need to caution them about their future.

10.3 POPULATION PRESSURE ON LAND AND AGRICULTURAL RESOURCES: GLOBAL CONTEXT

Population increase in many parts of the world has consequential effect on agricultural resource because an excessive growth of population can drastically minimize agricultural land throughout the world. It is reported that agricultural land which extracts food and cereals contain only 12% of the total land area of the planet which does not seem to be sufficient in terms of covering the subsistence of such a huge incumbent population. Of the remaining total 24% are arid grass land which is used for pasturing and grazing purposes, and another 30% is covered by forest necessary to protect the environment from greenhouse effect and other climatological imbalances. The remaining 34 per cent of the total land of the planet is fully unusable for any crop production as they are stony, too steepy or are exposed to extremely dry, cold and wet atmospheric conditions (Buringh, 1989). These lands are simply geologically infertile, unusable for pastures as grass land, and climatically unsuitable for crop production (Pimentel, 1989).

Thus, it becomes logical that when population grows at an unlimited rate, it obviously puts pressure on our marginally available 12% of useable agricultural land, the supply of which is also shrinking day by day (Note 4). An extreme growth of population also squeezes the per capita availability of cultivable land. Based on evidence, it is calculated that at present, we require 0.5 hectare per capita crop land as a minimal requirement to sustain a proper diet and nutrition. But due to continuous population growth and also rapid land degradation, the availability of per capita land is reduced to an extreme point day by day (Leach, 1995). In many Third World countries, it is far below the global average, putting people under serious food shortage and effectual causation of poverty and hunger. An example can be found in Honduras where John Pender (1999) formulated a hypothesis on the impact of rural population on productivity, poverty and natural resource management. Pender's argument seems to be relevant for this research which also has shown the implications of demographic pressure on the farm-based agrarian households in rural Bangladesh. There is however, another research conducted by D.G. Satihal, L.D. Vaikunthe and P.K. Bhargava (n. d.) which documents those rapid demographic and agricultural changes in various parts of Karanataka District for the last few decades. The paper has however, shown that there is a large variation in the general land utilization pattern and availability of cultivated land in different parts of the district. Bivariate analysis of data however, suggests that agricultural growth in all cropped areas of the district largely lag behind population growth, except in a few areas where there are higher growths of food crops.

We know that land and its terrestrial environment is essentially an important natural resource which provides 99% of humans' food requirement (Pimentel & Pimentel, 1996). Thus logically, it is quite likely that when this land is under serious threat due to population growth, farmers need to use the same land repeatedly through intensive multi-cropping production. When farmers go for intensive cultivation, they have to utilize mechanized farming and make an abrupt shift from their traditional indigenous farming system. The introduction of mechanized farming provides a sharp increase of crop production which is essential to support a growing population. Traditional subsistence farming in Asia and Africa in the past involved the rotating cultivation or mono-cropping, keeping the land fallow for some time, which as a matter of fact allowed the land to be revitalized and regain its nutrients. But with the increase of population, people put continuous pressure on land, without allowing them any time off. The resulting consequence is the deterioration of the soil which keeps the land fully dependent on chemical fertilizer and uncontrolled irrigation. Therefore, peasants moving towards mechanized farming no longer depend on seasonal rain and also at the same time, are totally dislodged from indigenous farming mechanisms. Due to mechanized farming, crop production increases, yet a complimentary notion develops when people usually care less about reducing the population.

10.4 DEMOGRAPHIC IMPACT ON AGRICULTURAL LAND USE: A MICRO-LEVEL EXAMPLE FROM AN ETHNOGRAPHIC RESEARCH

This part of the research has its ethnographic documentation on two villages in Bangladesh. Dhonjoypara and Gopalhati, are both agricultural communities located in the same physiographic and environmental setting. They belong to Puthia union (Note 5) of Rajshahi District in the north-western part of Bangladesh. Physiographically, Puthia and these villages lie on the outer margin of the riparian tract which is about eight miles in land from the left bank of the Padma River. It lies in the southern low-lying bed or depressed marshy area (Siddique, 1976). It is situated over 25°22 north latitude and 88°50 east longitude (Hossain et al., n. d.). The mean temperature for Puthia and these study villages increases from 63° F in January to more than 85° F in the summer months. Of the yearly rainfall of about 56 inches, no less than 50 inches fall in the rainy season. Compared to other parts of the country, the rainfall of Puthia villages is far less, which speaks of the necessity for irrigation of its land.

On the basis of surface level, there are three types of land in Puthia (Note 6): (1) Daira or also known as bhiti land, meaning land for homestead or the elevated land above flood-level; (2) Mathan or the flat fields of intermediate level which are partially flooded during the rainy season; and (3) Layal or the low-lying land which is completely flooded during the rainy season. These diverse soils of the villages provide them with diversified cropping pattern (Note 7).

10.5 SETTLEMENT TRENDS: AGRICULTURAL LAND USE AND THE POPULATION DYNAMICS IN THE VILLAGES

We will examine the settlement trend, land use pattern and the population dynamics in Dhononjoypara and Gopalhati to provide data for the micro-level investigation. Information about village settlement prior to 1850 is not available. According to the first Village Revenue Survey of 1850, Dhononjoypara contained 205 acres of land, while Gopalhati had 393 acres of land in its mouza. The Village Revenue Survey of 1850 indicated that mouza Dhononjoypara at that time had only five households occupying a total of nine acres for homesteads. The amount of cultivable land in Dhononjoypara was 190 acres and the remaining 6 acres were waste and uncultivated. On the other hand, Gopalhati had 14 households having 25 acres for homesteads; the amount of cultivable land was 368 acres.

TABLE 2: Land-use pattern for villages Dhononjoypara and Gopalhati villages since 1850 (in acres)

	Dhononjoypara				Gopalhati		
Census Year	Land Used for Settlement	Land Used for Cultivation	Waste and Uncultivable Land	Total	Land Used for Settlement	Land Used for Cultivation	Total
1850	9	190	6	205	25	369	393
1968	58	152	--	210	99	293	392
1974	No information	No information	No information	210	No information	No information	399

The Census Reports of 1951 and 1961 provided information on population, household and literacy but did not give any information on settlement pattern, thereby making it impossible to analyze in detail the changes over time. However, the Revisional Settlement Survey of 1968 which came out in 1978 filled the vacuum in this regard, and the Census Report of 1974 provided gross data on the total amount of land available in each village. According to the Revisional Settlement Survey (1968), mouza Dhononjoypara had 210 acres of total land of which 152 acres were cultivated. In Gopalhati, there were 392 acres of total land, of which 99 acres had been used as settlement and the remaining 293 acres were agricultural land. It may be inferred from the Revisional Settlement data that there has been a tremendous increase of land for settlement in both villages since 1850. The reason is obviously the increase in population.

In addition to natural growth of the local population, in-migration to this low density populated area from the high density districts of eastern Bangladesh further augmented the population growth. Khan's (1977) study of migration in Puthia villages can be cited as an example to support this statement. The Census Report of 1974 estimated 210 acres of total land for Dhononjoypara, and 399 acres for Gopalhati. This estimate for Dhononjoypara is quite consistent with the Revisional Settlement data, but for Gopalhati, the figure projects a slight variation in the estimation, the reason remains unknown to us (Table 3).

TABLE 3: Demographic data for villages Dhononjoypara and Gopalhati (selected years 1951-2011)

Year	Village Dhononjoypara					Village Gopalhati				
	Number of HH	Total Population	Male	Female	Literacy	Number of HH	Total Population	Male	Female	Literacy
1951	No data	350	No data	No data	No data	90	380	No data	No data	No data
1961	72	371	195	176	15.36	93	513	264	249	11.11
1974	62	348	164	84	2069	151	959	478	481	11.57
1985	105	660	345	316	23.60	196	1207	631	576	14.17
2011	199	915	471	444	43.60	299	1964	997	967	44.50

Source: Census reports 1957, 1961, 1974 and field study 1984-85; field reports, 2011.

According to the village census that was administered during the field work (August 1984 to August 1985), there were 660 people in Dhononjoypara, of which 591 (89.55%) were Muslims and the remaining 69 (10.45%) were from a tribal community named the Santal. In Gopalhati, out of a total 1207 people, Muslims constituted 1142 (94.61%) while 65 (5.39%) were the Hindus. The average household sizes for Dhononjoypara and Gopalhati were 6.28 and 6.18 persons respectively which were slightly higher than the national average. Data gathered from my revisitation of the villages indicate that there has been a sharp increase in the population in recent years and the statistics show that within a span of 25 years, the increase rate of population growth in both villages has gone up to 80% and it is suspected that it will take only 4-5 years to double the percentages (Table 3).

To have a clear picture of population growth in Dhononjoypara and Gopalhati, a demographic view of the villages since 1951 to present time is shown in Table 3. To enquire about the population transition of the villages for the past century it is necessary to know the population dynamics of Puthia Union and Puthia Upazila as a whole. Census recording in the sub-continent of Bangladesh-India and Pakistan began as late as 1872. But there is no information on population at the village level, nor does the Village Census of 1901 dealt specifically with village statistics. It was simply a camouflage in the name of Village Census. In fact, population statistics at the village level only came into existence in 1951.

The reported census of 1872 and 1901 produced data on the thana (i.e., present upazila) level which indicate that there has been decades of declining population in Puthia Upazila. The Bengal District Gazetters-Rajshahi-1916 (O'Malley, 1916) indicate that the population of Puthia and adjoining Upazilas (i.e., Bagmara, Mohanpur, Paba & Charghat) declined tremendously between 1872 and 1891 due to prevalence of malaria, smallpox, and cholera together with water-hyacinth which blocked the water channels. This caused a 15.01% decrease of population growth (1976, p.48). Through 1901, Puthia and the adjoining upazilas sustained a loss of population by a decrease of 12.08 % (O'Malley, 1916). Many people died in this swampy water-logged area and others migrated to the comparatively healthier and more prosperous areas (Naogaon & Panchupur, n. d.) of Rajshahi Region (O'Malley, 1916). Nelson (1923) reported that the population of Puthia decreased by 44% between 1872 to 1912. This declining population trend for Puthia continued until 1951 due to a large emigration of the Hindus to India during and after 1947 (Hossain et al., n. d.). From 1951 onwards, the population had again increased in Puthia Union, as it had for Dhononjoypara and Gopalhati. The increase of population for Puthia Union between 1951 and 1960 is 32.2% (Hossain et al., n. d.). This growth rate has been mostly due to increasing birth rates. Side by side with the population growth, we find that there had occurred a tremendous loss of the agricultural land in the villages of Dhononjoypara and Gopalhati.

10.6 FINDINGS AND DISCUSSIONS

Based on our foregoing discussion, it is discernible that since the beginning of human history, global population had increased tremendously at a continued process putting the people on the earth in enormous problems and economic hardship. It has been evidenced that when the world population was 1 billion until 1830, it took only 100 years to double the number to be 2 billion in 1930. We found that within a span of 45 years, this number has reached to another double to make it 4 billion in 1975. There was an addition of another two billion people by the end of the Twentieth century to enhance world population 6 billion in 1999. Thus it has been observed that with the passing of time in every sequence, doubling time for population increase is becoming lesser and lesser. Such increase of population requires more and more land for settlement and habitation having a direct impact on natural resources. It has been calculated that overpopulation is the prime reason for reducing per capita availability of land. To substantiate this statement, the paper has incorporated an ethnographic documentation of two villages from Bangladesh where it has been evidenced that demographic pressure has resulted in the transformation of agricultural land. Based on the data, it has been found that this loss of land in the villages Dhononjoypara was 18% and for Gopalhati, it was 19%. During my field-visits in early 2012, the villagers reported to me that almost half of the agricultural land in both the villages has been taken for settlement until recently. It is thus indicative that population increase puts heavy pressure on cropland when the people in the rural areas have been compelled to divert their farming land for purpose of settlement and habitation.

10.7 CONCLUSION AND RECOMMENDATIONS

Rapid population growth has been identified as the single most important factor for environmental degradation, which puts people in extreme poverty and also deteriorates human life in many nations of the world. It causes tremendous transformation of the world's natural landscape to agriculture, and it has been documented clearly in our discussion that agricultural resource will soon exceed its capacity by making an irreversible damage to the ecosystem of this planet. In my paper, I have documented the gradual increment rate of population growth in the global context, and also at the same time, I have provided ethnographic documentation of the pattern of such growth at the village level. It has been argued by a few scholars (e.g., Buchholz, 1993; Karim, 2010) that the future of food security and land protection entirely depends on the control of and restriction on births than emphasis on unusual use of chemical fertilizer to boost agricultural production. There is argument that it is more humane and ethical to implement family planning programme to keep the population at the lowest level than allow people to be victims of starvation. To allow the increase in population is nothing but to cause starvation, health problems, increasing unemployment and finally the destruction of the environment. This is particularly true for some Third World countries like India, Pakistan, Bangladesh, Ethiopia, Indonesia and Nigeria where there is rapid increase of population among the lower-income and poorer sections. It is assumed that family planning programme do not work properly in these countries at the rural level. Government s and the NGOs in many of these countries try to popularize family planning programme among the wealthy, rich and educated segments of society who are nevertheless are quite aware of the situation (Karim, 2010). We must remember that when population increases effectually much cropland is taken for urban and rural habitation. In order to feed many mouths, farmers have to use excessive fertilizers and pesticides which eventually destroy the fertility of the soil and also at the same time, negatively affect human health. It has been proven from the ethnographic examples given in this paper that there has been a tremendous shrinking of agricultural land in the two villages of Bangladesh within the span of one hundred years. This is particularly true for Bangladesh as well as for other developing nations like India, China and Indonesia, which have all the potential of development but often lag behind because of their demographic pressure and man-land low achievement (see Kumar, 2000 for details on this).

Based on the above contention, however, I formulate in my conclusive statement a very simple formula below for the peasants around the world in regard to their agricultural land-use pattern throughout the years, stating that, we must find out a mechanism to deal with such a huge number of growing population around the world at this critical situation. The first suggestion is that each and every nation around the world should formulate their own policies immediately with a target to bring the population growth at a replacement level, and simultaneously they should invent some techniques to solve the food crisis throughout the world.

NOTES

1. In terms of availability of natural resources, this planet has already exceeded its upper limit of the environmental carrying capacity. When this carrying capacity exceeds, there is an irreversible damage on the ecosystem. The overuse of land and depletion of land resources are the best examples to show the consequence of exceeded carrying capacity (see Pimentel, 1998 for details).
2. Global estimates of tropical deforestation range from 69,000 km in 1980 to 10000 - 165000 km. in the late 1980s (see David & Tucker, 1997; for details also see Tucker & Richard 'Global Disforestation in the Nineteenth Century World Economy', 1983; Durham & Williams, Prog Human Geography 13, [1989]).
3. Thomas Robert Malthus is perhaps the first theoretician who propounded demographic theory with a pessimism showing that population increase will supersede food supply and resources which create problems for human survival [see Malthus, 1798].
4. During the 1930s, the number of population was lowered and the reason was the decline in fertility, and the higher rate of mortality due to a number of reasons.
5. A union is the lowest unit of local government organization in Bangladesh. It usually comprises of 6-15 villages and is governed by one chairman and nine members directly elected by the villagers.
6. Physiographically, Rajshahi District is divided into three broad divisions: (a) the Barind Region, (b) the newly laid alluvial deposits along the bank of Padma and (c) the beel or marshy area (see Karim, 1990 for details).
7. I have taken a few sample families from different class of peasants to represent their groups. In future, this will further be updated by interviewing each family to know specifically their everyday food consumption patterns.

REFERENCES

1. Ahmed, A., & Ryosuki, S. (2000). Climate Change and agricultural food production at Bangladesh: an impact assessment using GIS-based Biophysical Simulation

TABLE 4: A Suggestive Model for Reverting to the Indigenous Farming System

Total Suggested Use of Cultivable Land	Farming Method	Purpose(s)		Plans Regarding Population in the Global Context
Fifty percent (50%) of the total cultivable land	Modern intensive cultivation	Scientific cultivation	Diverse food production with huge amount of HYV rice production	a) Population in the third world and poor countries must be kept at the replacement level. b) There should be controlled population policies in many rich and developed nations. c) Controlled population in many Asian and Middle Eastern countries.
Twenty-five percent (25%) of the total cultivable land	Mono-cropping or double cropping	Indigenous cultivation	Indigenous food production as indigenous technology permits	a) Population in the third world countries and many selected underdeveloped countries must be brought at the zero level replacement. b) Optimum population in many rich and developed nations (optimum population for many Asian and Middle Eastern countries as well as East Asian countries).
Twenty-five percent (25%) of the cultivable land	Keeping it fallow for the whole year	To maintain land fertility and nutrition		a) Population in the third world and many selected countries must be at zero level replacement. b) Controlled population for many rich and developed nations. c) Controlled population in many rich Asian and Middle Eastern countries.

model. A paper published by the Center of Spatial Information Science in Tokyo, Japan.
2. Ahmed, I. (1979). Employment in Bangladesh: Problems and Prospects. A paper presented at the Fourth Economics Association Conference. Dhaka, January 6.
3. Buchholz, R. A. (1993). Principles of Environmental Management. New Jersey: Prentice-Hall.
4. Buringh, P. (1989). Availability of agriculture land for crop and livestock Production. In D. Pimentel & C. W. Hall (Eds.), Food and Natural Resources (pp. 69-83). San Diego: Academic Press.
5. Cuffaro, N. (1997). Population Growth and Agriculture in Poor Countries: A Review of Theoretical Issues and Empirical Evidences. World Development, 25(7), 1151-1163. http://dx.doi.org/10.1016/S0305-750X(97)00025-9
6. Foley, J. A. (2005). World land use seen as Top Environmental issue. Science Daily. Special Bulletin. Government of Pakistan. (1951). Census of East Pakistan. Karachi: Government Press.
7. Government of the People's Republic of Bangladesh. (1974). Village Census (1974). Dhaka: Bureau of Statistics.
8. Hossain et al. (n. d.). The Pattern of a Peasant Economy, Puthia-A Case Study. Rajshahi: Socio-Economic Survey Project, Rajshahi University.
9. Jaya, K., & Stanley, G. (2000). Population and Environment by 2000 AD - A Social Manifesto. The International Journal of Sociology and Social Policy.
10. Karim, A. H. M. Z. (1990). The Pattern of Rural leadership in an Agrarian Society: A Case Study of the Changing Power Structure in Bangladesh. New Delhi: Northern Book Center.
11. Karim, A. H. M. Z. (2011). Agro-based Food Production System in Bangladesh: A Socio- Demographic Impact Assessment from Asian Examples. The Social Sciences, 6(6), 473-479.
12. Khan, S. H. (1977). Beyond the Trap of Traditions. In Migrants and Locals in a Rural Community of Bangladesh. Unpublished M. Phil. Thesis. Institute of Bangladesh Studies, Rajshahi University.
13. Lal, R. (1989). Land Degradation and its Impact on Food and other Resources. In D. Piementel (Ed.), Food and Natural Resources. San Diego: Academic press.
14. Leach. (1995). Global Land and Food in the 21st Century. Stockholm: International Institute for Environmental Technology and Management.
15. Malthus, T. R. (1798). Essay on the Principle of Population. London: Macmillan.
16. Meyer, B. W., & Turner, B. L. (1992). Human Population Growth and Global Land-Use/Cover Change. Annual Reviews Ecological Systems, 23, 39-61. http://dx.doi.org/10.1146/annurev.es.23.110192.000351
17. Myers, N. (1993). Gaia: An Atlas of Planet Management. Garden City: Anchor and Doubleday.
18. Nelson, W. H. (1923). Final Report on the Survey and Settlement Operation in the District of Rajshahi: 1912-1922. Calcutta: The Bengal Secretariat Book Department.
19. O'Malley, L. S. S. (1916). Bengal District Gazetteers: Rajshahi. Calcutta: The Bengal secretariat press.
20. Pender, J. (1999). Rural Population Growth, Agricultural Change and Natural Resource Management in Developing Countries: A Review of Hypotheses and some

evidences from Honduras. EPTD Discussion paper. Washington D.C. International Food Policy Research Institute.

21. Piementel et al. (1994). Natural Resources and Optimum Human Population. Population and Environment, 15, 1117-1123. http://dx.doi.org/10.1126/science.267.5201.1117
22. Piementel et al. (1996). Food Energy and Society. Colorado: University Press of Colorado.
23. Pimental, D. (1997). Environmental and Economic costs of Soil. In U. Tim (Ed.), Environmental management: readings and case studies. Oxford: Blackwell Publishers.
24. Pimentel et al. (1995). Environmental and economic cost of soil erosion and conservation benefits. Science, 267, 1117-1123.
25. Pimentel et al. (1998). 'Impact of a Growing Population on Natural Resource Management: The Challenge for Environmental Management'. In B. Nath (Ed.), Environmental Management in Practice (pp. 6-21). London and New York: Routledge.
26. Pimentel, D. (1989). Ecological Systems, natural resources, and food supplies. In D. Pimentel & C. W. Hall (Eds.), Food and Natural Resources (pp. 1-29). San Diego: Academic Press.
27. Pimentel, D. (1993). World Soil Erosion and Conservation. Cambridge: Cambridge University Press. http://dx.doi.org/10.1017/CBO9780511735394
28. Pimentel, D., & Marcia, P. (2006). Global Environmental Resources versus World Population Growth. Ecological Economics, 59. http://dx.doi.org/10.1016/j.ecolecon.2005.11.034
29. Qu, G., & Li, J. (1992). Population and Environment in China. Beijing: China Environmental Science Press (In Chinese), referred in Pimentel (1994).
30. Ramchandan et al. (2008). Population Growth: Trends, Projections, Challenges and Opportunities. Sathial, D. G., Vaikunthe, L. D., & Bhargava, P. K. (Eds.). (n. d.). Population Growth on Agricultural Land Utilization in Karnataka, India. Research Centre. JSS, Institute of Economic Research, Karnataka University Dharwad.
31. Schneider et al. (2011). Impacts of population growth, economic development, and technical change on global food production and consumption. Agricultural Systems, 104, 204-215. http://dx.doi.org/10.1016/j.agsy.2010.11.003
32. Siddique, A. (1976). Bangladesh District Gazetter-Rajshahi. Dhaka: Bangladesh Government Press.
33. Skole, D., & Comton, T. (1997). Tropical deforestation and Habitat Fragmentation in the Amazon: Satellite Data from 1978 to 1988. In A. O. Lewis & U. Tim (Eds.), Environmental management: readings and case studies. Oxford: Blackwell publishers.
34. Timah, E. A., Nji, A., Divine, F. T., Leonard, M. N., & Irne, B. B. (2008). Demographic pressure and natural resources conservation. Ecological Economics, 64, 475-483. http://dx.doi.org/10.1016/j.ecolecon.2007.08.024
35. UNDP (United Nations Development Program). (2009). Human Development report 2009. New York: Oxford University Press.
36. United Nations Population Reference Division. (2009). World Population Prospects 2007. Retrieved February 9, 2009, from http://www.data.un.org
37. Wright, R. (2008). Environmental Science: Toward a Sustainable Future. NJ: Pearson: Prentice Hall.

CHAPTER 11

CROP BREEDING FOR LOW-INPUT AGRICULTURE: A SUSTAINABLE RESPONSE TO FEED A GROWING WORLD POPULATION

TIFFANY L. FESS, JAMES B. KOTCON, AND VAGNER A. BENEDITO

11.1 INTRODUCTION

As the world population increases and the availability of resources decreases (Figure 1), the need for efficient food production has become paramount (Table 1). Modern high performance varieties are usually bred for high-input systems. However, as resources decline and populations grow, high-input systems become less sustainable and realistic. In the future, maintaining high input systems will become increasingly difficult due to reductions in the availability of required resources, such as water, oil, and phosphorus. We acknowledge that there are numerous social and economic issues (poverty, illiteracy, disease, politics) around the world that contribute to the low productivity of regional cropping systems as well as improvements in production management, and post-harvest handling and storage that could be improved upon to decrease the pressures associated with feeding a swelling population. However for the purpose of this

This chapter was originally published under the Creative Commons Attribution License. Fess TL, Kotcon JB, and Benedito VA. Crop Breeding for Low Input Agriculture: A Sustainable Response to Feed a Growing World Population. Sustainability **2011,**3 (2011); pp. 1742-1772. doi:10.3390/su3101742.

review, focus will be concentrated on the technical aspects of breeding that accommodate future food demands in a world of decreasing resource availability. By using more energy-effective approaches to breeding, varieties can be developed that are best suited to specific agricultural ecosystems, allowing for maximum production in that particular settings. Plant breeding programs focused on developing genotypes adapted to specific agricultural environments and lower inputs could help attain sustainable, higher productions with lower energy costs to accommodate the growing population, while providing an adequate food supply and responsibly managing declining resources.

TABLE 1: Past, current, and projected future population sizes, along with changes in food production and resource consumption.

Production Demands	Year		
	1960	2000	2050
Population (billions)	3	6	8.7–10
Food production (Mt)	1.8×10^9	3.5×10^9	6.5×10^9
Agricultural water (km^{-3})	1500	7130	12-13,500
N fertilizer use (Tg)	12	88	120
P fertilizer use (Tg)	11	40	55–60
Pesticide use (Tg, active ingredient)	1.0	3.7	10.1

Mt, Metric ton; km^{-3}, cubic kilometer; Tg, 10^{12} g or million metric tons. [1,2,5-11].

11.2 MODERN AGRICULTURE AND BREEDING STRATEGIES: HIGH-INPUT PRODUCTION

In developed nations, modern agriculture is based on high-input agricultural systems, which is not sustainable given resource limitations projected to occur in the near future. High-input production systems often consist of large acreage monocultures relying on heavy machinery, high-yielding

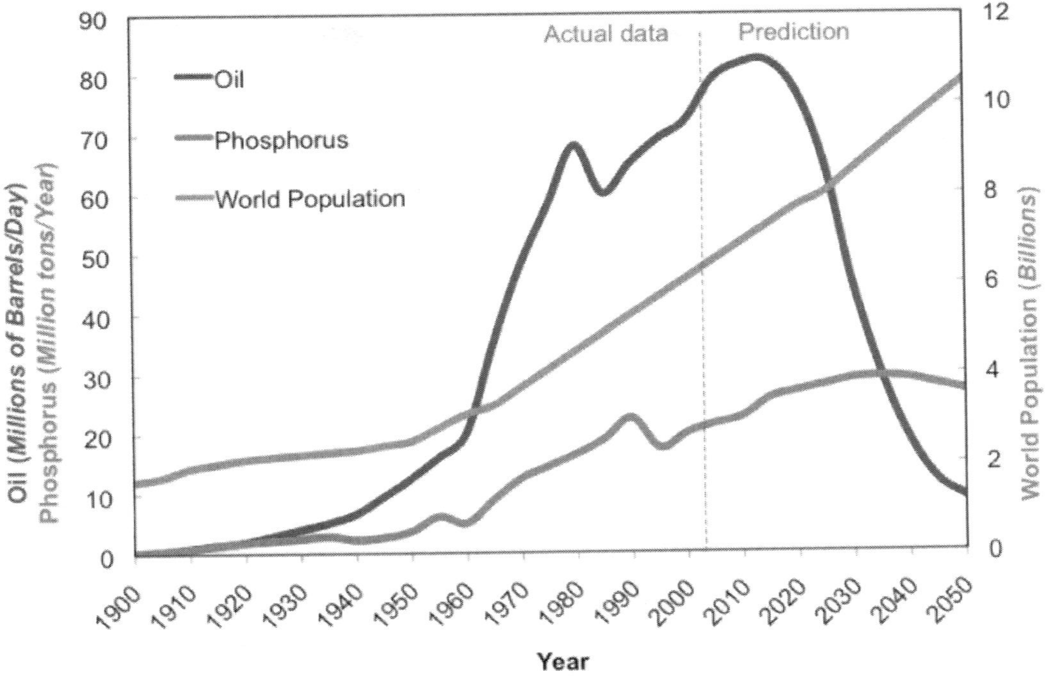

FIGURE 1: Expected population growth in comparison t resource availability. Situation of the so-called 'peak society' highlights the urgency of breeding crops for low-input systems and improved resource management, as population and food demands are expected to increase while global resources decline [1-4].

varieties, synthetic and natural fertilizers, frequent pesticide applications, and the use of irrigation. Developed mainly during the Green Revolution, modern high-input agriculture provided new and convenient farming practices allowing for adequate production, significantly reducing world famine and malnutrition. The focus of this type of agricultural system has been to create an environment that maximizes productivity and profitability, along with providing a relatively inexpensive food supply.

Although high-input systems may provide large yields, they create a fundamentally unsustainable environment that requires frequent and heavy applications of water, nutrients, and disease/pest/weed controls. In modern breeding programs, varieties are often developed by crossing parental genotypes possessing the most desired traits (e.g., yield, early flowering, vigor, plant architecture), and selecting offspring are under optimal growth conditions. The most successful individual plants are selected in successive generations to ensure consistent uniformity [12], in the case of self-pollinating crops (e.g., wheat, potato, pea). Even in outcrossing crop species (e.g., maize, canola, cotton), for which heterozygosity confers advantage through heterosis at the individual level, genetic variability is restricted in the breeding program pipeline by using only a few elite highly-inbred parental lines generated from distinct plant populations that are finally combined to produce genetically homogenous hybrids. Modern crop improvement programs generally select under optimal conditions, therefore the focus is on genotypic selection based on increased yield performance or fruit/grain weight. This method of artificial selection results with a predictably uniform crop, in which genetic variability is restricted. Due to the field conditions provided by high-input production system, this breeding regime has been the dominant approach during the last century. Genotypes selected for high performance in high-input conditions likely do not maintain those same high yields under low-input or stress conditions due to the lack of natural genetic variation [13]. High-input systems work mainly for producers in the developed world, where the heavy importation of supplies and governmental incentives guarantee production and competition. However, many of the food production systems around the world are either low-input or under stress conditions and cannot depend on the purchase of supplies or fiscal incentives for crop production, subsequently plant breeding programs for these systems have been largely neglected.

11.3 CURRENT WORLD SITUATION

11.3.1 POPULATION

The current population is quickly approaching seven billion people worldwide and is projected to reach between 8.7–10 billion people by the year 2050, an increase of 45% [5,14,15] (notice that the term 'billion' is used in the sense of the short scale system throughout the text: 1 billion = 10^9). The rate of population growth varies around the world, however the greatest increase in population growth is expected to occur in developing and poverty-stricken countries concentrated in Africa [16]. Currently, the global population is increasing at 1.1% per year [4], and although there has been a deceleration over the past few decades, the overall growth in population has been positive. This population increase, which has been continuous since the Bubonic Plaque (1338–1351), will result in greater demand for food and agricultural commodities, while land and resources available for crop production will be on the decline. Accommodating the growing demand for food will undoubtedly be difficult, the UN Food and Agricultural Organization [17] projects that crops/livestock demand will rise 40% by 2030, reaching 70% by 2050. More specifically, the increased demand for cereal grains (human and livestock feed) will require production to increase from the current annual production of 2.1 billion tons to 3 billion tons by 2050. The demand for animal protein products will be even greater. Production will need to increase 200 million tons by 2050 to meet the projected 470 million ton demand [5]. In order to meet this demand, agricultural production will have to be as efficient as possible, especially in low-yielding agroecosystems. The potential for the human population to exceed resource availability, and therefore the carrying capacity of the environment is realistic. With proper and efficient breeding technologies that address low-input conditions, varieties that are geared toward limited or stressed agroecosystems could alleviate the production pressures associated with population increases.

11.3.2 LAND

With the increasing population and consequent food demand, proper resource management will be essential in creating a balance between human activities and environmental sustainability. One of the many global concerns with the ability to produce sufficient food for the growing population is the availability of arable land. According to FAO, there is sufficient land space to feed future global populations. Currently, 1.6 billion hectares of land are used for agricultural purposes (almost the size of Russia), but FAO estimates that there is as much as 2.4 billion hectares suited for agricultural expansion of wheat, rice, and maize cultivation [18-20]. However, when factoring in the importance of maintaining biodiversity and the carbon cycle, there is anywhere between 50 million–1.6 billion hectares potentially available for greater agricultural production [20]. In addition, much of the land often considered suitable for agricultural conversion has chemical and physical constraints, lack accessible roads, endemic diseases, or is covered by forest. Land available for agricultural expansion that is currently uncultivated, non-forested, relatively unpopulated, and not under government protection totals 455 million hectares, which are largely concentrated in Brazil, Argentina, and Sub-Saharan Africa [21]. The conversion of this land to agricultural production could cause severe environmental and social consequences [22]. Potential to reclaim agricultural land, approximately 26 million hectares, left abandoned after the collapse of the Soviet Union however does exist [22].

Ironically, population increase has been historically correlated to farmland loss despite of an increased demand for food. Over the past 40 years there have been significant farmland losses around the world, most notably in China, South Korea, India, and the United States. Between 1980 and 2000, the U.S. population grew by 24%, approximately 50 million people, during the same time 34% of arable and forestland was converted to accommodate urban sprawl [23]. Also during this time period, agricultural expansion in the tropics came at the sacrifice of both intact and disturbed forest [22]. The increasing trend in population density and subsequent urban development is projected to continue by FAO [5], with magnitude varying by region and competition with the energy sector [24]. In the U.S., the developed area is expected to swell by 79%, to occupy roughly

9% of the total land base [23]. Lambin and Meyfroidt [22] estimated that anywhere between 81–147 million hectares of additional cropland will be needed to produce food for the 2030 population. Although there are major uncertainties and inconsistencies involved with estimating the potential of land for agricultural purposes, it seems inevitable that the demand for land will progressively increase as the population and demand for food increases.

11.4 REALITY OF THE SITUATION IN MODERN AGRICULTURE

11.4.1 LOW-INPUT PRODUCTION

Low-input farming can be defined as systems managed with reduced use of inputs, usually resulting in a system that suffers from some type of limitation or stress, commonly nitrogen and phosphorus deficiencies or inadequate water supply, that ultimately cause yield losses. These systems are not necessarily organic in practice (as defined by the USDA), since conventional low-input and organic high-input operations are not unheard of around the world. However low-input is often associated and used as a synonym for organic production systems, especially in developed countries. Low-input systems have reduced use, but not elimination of fertilizers (either from inorganic or organic sources), or pesticides and herbicides (either biological, inorganic, or organic). Low-input systems rely on the improved management of on-farm resources, consequently resulting in a more sustainable agroecosystem, due to a reduced dependence on off-farm resources, including energy inputs such as gas and oil, in comparison to modern high-input systems.

One billion and four hundred million people in the world, mostly in developing nations, rely on crops grown in low-input systems as the primary source of agricultural production [13]. Low-input or resource-poor farmers account for half of the world's food producers, providing upwards of 20% of the global food supply [25]. Despite the high number of low-input producers globally, these resource-poor farmers have not benefited as much from modern breeding programs. This is largely thought to be due to varieties being developed under conditions not represented by marginal environments [26], which can be defined as areas with severe restricting factors (or access to means of alleviating it) for acceptable crop performance that are generally imposed by inherent local edapho-climatic conditions (or lack of technology to circumvent them). Unfortunately, varieties that are best suited to stress conditions are limited, inaccessible, or costly. Due to the nature of breeding varieties for modern agriculture, conducted largely under optimal high-yielding conditions, varieties that possess genetic traits advantageous in low-input systems are often overlooked [27].

Several recent studies have shown that modern varieties can be out produced by traditional farm varieties under low-yielding conditions (Figure 2). Toure et al. [28] found that under low management, traditional lowland rice varieties yielded more than modern varieties in Africa. Similar results have been observed with other major crops, such as sorghum, wheat, and barley [26,29,30], however this is not the case when produced in mediate or high-input systems. Modern breeding programs have some success at releasing maize and wheat varieties intended for marginal production areas [31,32], however for the most part modern varieties have not been widely accepted by local farmers. In India, improved varieties of upland rice with yield advantages are available, however most farmers use traditional production practices that are incompatible, and therefore adoption of these varieties has been slow [33]. Marginal farmers in Ethiopia have also failed to adopt varieties recommended by breeders for the region due to their inability to adapt to the variety soil conditions present [26]. Modern varieties are often developed by crosses between exotic lines that are unable to adapt to stress or limiting conditions, as seen in several locally unaccepted sorghum varieties [34].

Modern varieties that have been successfully adopted by low-input producers generally have been developed using local germplasm, increasing genotype × environment (G × E) interaction, adaptability, and therefore crop performance [35]. It has been shown that the most efficient way to improve yields under low-input conditions is to select varieties while under low-input or stress conditions [36]. However, this practice is done by few breeding programs, leaving low-input producers without suitable cultivars. Low-input farmers in developing regions must rely on landraces or creole varieties, which have been selected for by the evolutionary process (naturally) and the local farmers (artificially), and are often exchanged among them.

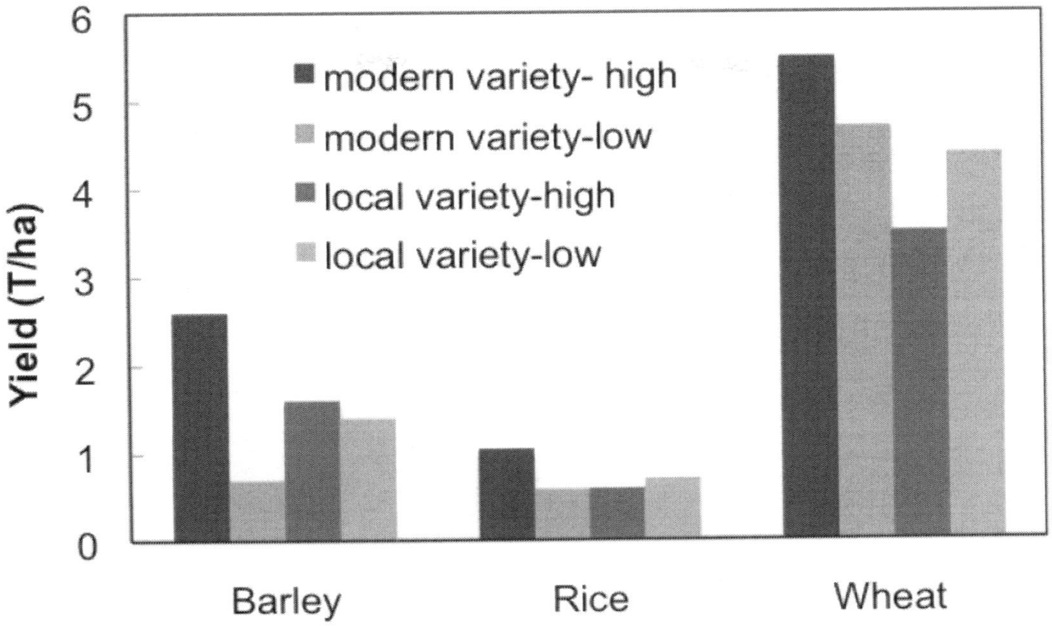

FIGURE 2: Yield comparison of commercial and local varieties produced under high and low-input conditions. This illustrates how system management and production practices influence yields of different crops. It is important to note that the influence of production practices, in terms of yield differs among crop varieties, indicating the significance of breeding programs that focus on differences in production practices and system limitations. Commercial varieties tested were 'Shege', 'Anjali', and 'Abbondonza' for barley, rice, wheat, respectively, while local varieties tested 'Himbil', 'Brown Gora', and 'Colognal Veneta' [26,30,33].

Having several definitions over the history of crop production, landraces are heterogeneous crop populations possessing genotypes specific to a given region, with great adaptability to the natural environment and agricultural practices of that region [37]. These varieties generally have not been optimized for yield performance alone, but are known to have higher yield stability and when produced under local stressed conditions, they are able to cope, producing moderate yields compared to modern cultivars that are unable to tolerate certain stresses and result in crop failure [37,38]. Breeding programs thus need to be developed that examine potential varieties more suited to low-yielding conditions, in which varieties would be selected that have more advantageous adaptations in stress conditions such as delayed leaf senescence, improved nutrient economy, local environmental fitness, consistent yield, and pest/disease resistance, thus increasing the profitability of sustainable low-input systems. The importance in shifting the paradigm of modern agriculture from high- to low-input is becoming more urgent as the human population continues to increase, at the same time crucial finite resources have or are reaching peak production and will inevitably begin to decline. Breeding for low-yielding and variable stress conditions is more complex than breeding for uniform, controlled, highly productive systems, but absolutely necessary to feed the growing population under diminishing global resources.

11.4.2 WATER

Besides the concerns of producing food for the growing population on land that is increasingly becoming more limited, we are also at a historical moment as the supply of available fresh water, oil, and phosphorus are reaching their peaks, all of which are key elements of modern production systems. Agriculture is the largest consumer of water worldwide, accounting for 70% of the global demand for fresh water. Currently, 1.2 billion people live and produce food in areas affected by drought, and this number is expected to rise as the demand for water increases [20]. Water demand is a function of several aspects including population density, diet, and agricultural practices of any given region. Thus as the population increases, the demand for water will also grow. With the projected increase in population density, water consumption is expected to increase 35–60% over the next several decades [20] and the agricultural demand for water will be competing with the increased demands of the industry and domestic sectors [6]. At present, 8% of the population, primarily in West Asia and North Africa (Libya, Egypt, Saudi Arabia, Iran, Iraq, Pakistan, and Afghanistan), and in South Africa are under intense drought conditions where water is the major constraint in food production [39]. Several countries are entering water shortages including populated regions like India and China, as well as Ghana, Ethiopia, Somalia, Kenya, and Zaire, which accounts for approximately 7% of the world population that must improve water access and utilization efficiency to meet future water requirements. Analyses have shown that the water demand for agricultural purposes can be improved by increasing the efficiency and expansion of irrigation systems [40]. In addition, breeding for characteristics that limit water loss and improve water use efficiency in crop plants will also significantly improve the ability of the water-deprived societies to provide food for their growing populations.

To provide varieties that are better suited for drought conditions, a host of molecular and physiological adaptations that improve water use efficiency can be selected for, such as superior hormonal physiology, increase in stomatal conductance, osmotic adjustment, as well as improved root architecture, to achieve higher yields under dry conditions. One of the simplest ways to improve yield in water-deprived systems is to gain access to water reserves in deep soil by breeding for increased root depth and distribution. It has also been noticed that selecting varieties that put a greater portion of energy into reproductive organs over vegetative production increases the harvest index of grains [41]. Unfortunately, identifying genetic targets in respect to drought resistance has been difficult. Better estimates for yield under stress can be reached by identifying varieties in water stress environments that are free of undesirable traits. This method of selection will likely involve a genetic shift toward dehydration avoidance, resulting in advantageous physiological changes such as early flowering, decreased plant height, and leaf area [42]. In general, these cultivars may have a lowered yield potential, compared to modern well-watered varieties, but are better able to adapt given water stress, allowing for yield improvements in dry regions or where dry spells occur.

Breeding for other less obvious traits, such as abscisic acid (ABA) sensitivity and high rate of osmotic adjustment have been implicated for yield improvement in water-stressed crops. Several studies have shown that improved water use efficiency can be achieved through ABA hormone control [43-45]. ABA is a hormone involved in a plant's response to stress conditions, including drought. As water availability in the root zone declines, ABA is synthesized by root cells and translocated to the shoot, leading to changes in solute concentration of guard cell cytoplasm via regulating ion channel openings in the plasma membrane, finally resulting in decreased stomatal conductance. All of which translates to guard cell deflation and stomata closure, therefore leading to water conservation. The genetic control of ABA physiology involves its biosynthesis, storage, distribution (transport), cell perception (receptors), and signal transduction pathways (secondary messengers and gene activation). The best alleles of these genetic elements can be combined to generate genotypes that optimally cope with stresses. Transpiration can be reduced due to ABA over expression in several plant species, including important and valued crops such as tomato, cowpea, and common bean [43,46-48]. A decrease in transpiration can improve growth, water status, and turgor of crops produced in water-limited systems, allowing for potential yield gains [43,49]. To improve water use efficiency, genotypes that have increased sensitivity to or production of ABA along with improved osmotic adjustment have been suggested as a selection trait for yield increase under drought stress. Osmotic adjustment is a cellular adaptation that enhances dehydration tolerance and supports yield under water stress. Rapid osmotic adjustment has been correlated to sustained growth and yields in water-limited systems [42]. This occurs because osmotic adjustment helps to maintain high leaf water content and turgor, aiding the crop in the continuation of moderate transpiration and photosynthesis under reduced leaf water potential, allowing for cell stability and avoiding yield losses. Thus, by selectively breeding for traits, whether physiological or molecular, that are advantageous during drought conditions it is possible to increase yields harvested from rain-fed or low-water input systems.

Molecular analyses using crop and model plant species are elucidating how plants respond to drought stress and recovery by using genome-wide expression regulation approach [50]. These studies may lead to identification of key genetic elements (genes) and genetic variations (of coding or regulatory regions) that are beneficial to drought tolerance. *HARDY*, an ethylene-responsive transcription factor, is an example of a single gene identified from studies with the model species *Arabidopsis* that has been demonstrated in rice to confer enhanced drought (and salt) tolerance when a specific mutant allele is overexpressed [51]. *ESKIMO1* is another promising example of a major genetic player affecting water economy (as well as cold and salt tolerance), with lack-of-function *Arabidopsis* mutants show better fitness in drought conditions [52] by altering hydraulic conductivity in the vascular tissues and increasing ABA levels [53]. Natural genetic variation studies in crops [54,55] and model species [56] are also helping to unveil alleles conferring drought tolerance traits, which can be used as powerful tools on breeding programs via either traditional or molecular approaches.

11.4.3 ENERGY AND NITROGEN

Modern high-input agriculture is heavily reliant on energy, particularly in the form of petroleum products like gas and oil. This dependency will cause the energy demand of modern agriculture to increase approximately 45% over the next 20 years in order to supply food for the increasing population [20]. Although there are alternative forms of energy (solar, wind, hydro, biofuel) to reduce the dependence on gas and oil, intense modern agricultural systems require significant energy input. Large equipment powered by fossil fuels are vital to today's crop production in order to prepare, cultivate, and harvest the vast number of crops grown to provide an adequate and nutritious food supply to people around the world. Petroleum supply, based on the production data, tends to follow a bell-shaped curve. The amount of oil in any given region is finite and the production of petroleum is quickly reaching its peak (Hubbert's peak), coinciding with the midpoint in the depletion of the resource supply. The U.S. reached its peak domestic oil production in 1970, and has since been importing more oil than it is capable of producing [57]. As the world's oil production peaks, maintaining high-input agricultural production systems will become increasingly more difficult and less productive.

Energy input is not only needed for the large machinery that is utilized in soil preparation, cultivation, harvesting, and multiple applications of fertilizer and pesticides, but also for chemical synthesis and long-distance transport of supplies and products. The Haber-Bosch process, used in the production of nitrogen fertilizer, allows for hydrogen, from natural gas, to be combined with atmospheric nitrogen to produce ammonia. This reaction must be performed under high temperatures and pressure, and is therefore energy expensive in itself. One metric ton of nitrogen fertilizer requires 873 m^3 (35 million BTUs, British thermal units) of natural gas, consuming 3–5% of the total U.S. production annually [7]. Currently, 88 million tons of nitrogen fertilizer is applied to crops around the world, 13% consumed by the U.S. alone, an additional 40 million tons are estimated to meet the global food demand of future populations [7,58,59]. Peterson and Russelle [60] estimated that, with proper alfalfa-corn rotations, the systems in the Midwest U.S. could see a 25% decline in nitrogen fertilizer reliance without experiencing significant yield loss. Successful and improved management of on-farm resources concurrently with improvement in nitrogen uptake and use efficiency could significantly reduce the energy used in agriculture, increasing the sustainability and productivity of low-input systems.

Without the foreseeable capability of supplying the energy demands required for its production, high-input agriculture will no longer be efficient or reliable for many producers. The consequences of removing fossil fuels from modern high-input agriculture can be noted in both North Korea and Cuba. Both countries, similar in many ways (land size, geography, and political isolation), have populations that depended on high-input practices to produce food during the last half of the 20th century. North Korea has no domestic oil or gas production, therefore it had to be imported from the former USSR after the Korean War (1953) to feed the population, and crop yields increased with the increased availability of fossil fuels and modern agricultural practices [61]. Unfortunately, with the collapse of the Soviet Union in 1990, access to gas and oil immediately ended. Consequently, soil fertility declined rapidly, equipment could not operate, and yields quickly and drastically decreased. As a result, around 3 million people subsequently died from famine during 1995–1998 [61]. On the other hand, Cuba also suffered when its import supply ended with the collapse of the Soviet Union. As production declined by nearly 54% (1989–1994), the solution formulated by the government was to transform the country's methods of food production from a high-input monoculture to a low-input, more sustainable system in order to avoid starvation and famine [61]. Work animals replaced tractors, natural pest controls became preferred over chemical applications, and state incentives attracted workers, all of which contributed to the production of a sufficient and successful food supply. Many of these practices are still common management strategies used today. Both countries are examples of what can happen when fossil fuels are removed from the modern agricultural system, and shed sufficient evidence on the importance of proper resource management, and breeding for low-input systems given the inevitable decline in valuable worldwide resources.

Plants require nitrogen (N) for growth and optimal yield, and although it is one of the most abundant recyclable elements on Earth, it is often the most limiting resources in agricultural systems. In high-input systems, nitrogen is heavily applied as an ammonium salt derived from the Haber-Bosch process, however in low-input systems nitrogen is provided primarily by managing on-farm resources (compost, manure, legume rotation) and nitrogen cycling in the soil is driven by soil microbes, often leaving low-input and organic farming systems with limited nitrogen pools [62]. Unfortunately, most breeding programs select varieties where fertilizers are liberally applied to ensure maximum production at optimal conditions, and are not well adapted to low-input systems. By breeding for varieties that are more adapted to limiting conditions, improved nitrogen use can help increase yields obtained from low-input systems, reduce fertilizer production, and potentially not only maintaining the energy required for crop production, but reducing it, all while feeding a growing population.

Baligar et al. [58] estimated that the overall efficiency of applied nitrogen fertilizers is around 50%, and that improvements in uptake and utilization can greatly increase the efficiency of fertilizers. Several studies have shown that for several traits, genetic inheritance was different in crops produced under high- and low-N inputs [63,64], indicating that different genetic elements are responsible for responding to different inputs. Gallais and Coque [65] determined that under low-N input, genetic variation in nitrogen use efficiency (NUE) is more important to yield improvement than nitrogen uptake. A clear understanding of the genetic mechanisms and inheritance of NUE is lacking for low-input systems, since most mechanisms governing NUE have been studied in high-input production systems [62,66]. Several adaptations have been suggested for advanced NUE, such as improved root development and architecture, along with delayed leaf senescence, increased arbuscular mycorrhizal colonization or nitrogen fixing symbioses, and increased activity of specific enzymes [7,58,67]. Under relatively simple genetic control compared to other beneficial molecular and physiological traits, improvements in root structure, such as length and thickness, as well as density by increasing the production of root hairs and adventitious roots are efficient ways to improve the ability for crops to acquire and absorb soil nutrients [68].

Breeding for delayed leaf senescence, increased enzyme production, and symbiotic relationships are more difficult than for qualitative (Mendelian) traits, largely because they are more complex and under polygenic control. Nevertheless they can be subject to genetic improvement through either traditional or advanced breeding [7]. Delayed leaf senescence is especially related to the physiology of the hormone cytokinin, and its associated genes seems to be key to molecular breeding of this trait under low-input conditions [69,70]. Bertin and Gallais [63] showed the NUE was negatively correlated to leaf senescence at low-input. By delaying leaf senescence, a genetic gain in yield can be seen, due to the greater capacity to uptake nitrogen during fruit maturity [62]. Also by breeding for high chlorophyll content, delayed leaf senescence can be achieved, leading to increased nitrogen uptake. Grain yield has been positively correlated to chlorophyll content in low-N input systems, this correlation not being significant under heavy nitrogen application [71,72], has been ignored by breeding programs and serves as an example of the need for specific breeding at low-input systems to improve sustainability on a global-scale. Spano et al. [73] determined that "stay green" genotypes of wheat had increased maintenance of leaf chlorophyll, which lead to 10–12% increase in grain weight. It was also determined that as a result of the increased N-uptake, due to increased leaf area duration and chlorophyll content availability, "stay green" varieties had early development of adventitious roots and increased root density than senescent varieties also contributing to yield gains [73].

Increased activity of certain enzymes may help improve a crop's utilization of nutrients. Malate dehydrogenase (MDH) is the enzyme used in the biosynthesis of malate, which is a requirement for respiration in nitrogen-fixing bacteria. Overexpression of MDH in alfalfa and subsequent malate exudation by roots via specific membrane transporters may improve nitrogen availability by providing substrates needed for respiration of the rhizosphere microflora, resulting with an increase in nitrogen fixation, thus improving the availability of nutrients to crops [62]. Malate export to the soil comes as an energetic expense to the plant, since photosynthates are released to the rhizosphere, and this activity must be tightly controlled to be energetically and nutritionally beneficial to the plant. Glutamine synthase (GS) metabolizes glutamate to glutamine, with ammonia as a required substrate for this reaction. Gallais and Hirel [64] showed that GS activity is positively correlated to yield increases. GS may ultimately determine rate of translocation of stored nitrogen to developing fruits, resulting in yield increases under low-N conditions. Gallais and Coque [65] suggested that high GS activity is a mechanism used by crop plants to prevent embryo abortion, specifically in limited nitrogen systems, resulting in increased the potential yield.

Improved interaction with soil microorganisms can significantly increase the efficiency of nitrogen use in crop plants. Arbuscular mycorrhizal (AM) fungi have been shown to improve nutrient uptake and potential yield of crops, although most attention has been given to phosphorus uptake. In addition, nitrogen plant nutrition can also be improved via mycorrhization [74,75]. The advantages of mycorrhizal colonization are most observed when N-input is limited, implying that the symbiosis may not significantly aid in the uptake of nitrogen from fertilizer [62,75]. Colonization by AM fungi among wheat cultivars started to decline during the 1950's, coinciding with the increased availability and heavy application of synthetic fertilizers. In a study by Hetrick et al. [76] found that wild traditional farm varieties, or heirlooms by today's standards, had increased AM symbioses which subsequently led to better growth compared to modern varieties. There is also indication that selecting varieties under low-N conditions may inadvertently result in a genetic shift toward AM associations and improved use of soil nutrients [77]. As an extension of the root system, mycorrhizal association is also thought to improve drought tolerance [78]. Although demonstrated for some species [79,80], and key genes of symbiosis establishment revealed [81], the genetic inheritance of this association in crop species is still elusive and deserves more attention of biologists, geneticists and breeders. However by improving any of the molecular or physiological features that improve the ability of crops to access soil nitrogen, crop growth and yield under stress conditions could increase with concomitant reduction of the energy required to meet the growing food demand.

11.4.4 ENERGY AND PESTICIDES

The production, transportation, and application of pesticides are all energy expensive, consuming 15% of the energy resources used by agriculture [82]. Many pesticides are manufactured using ethylene and propylene, both of which are made from catalytic reactions with crude petroleum or methane produced from natural gas. The production of modern pesticides consumes between 2,000–6,000 BTUs per kilogram of material, depending on the final chemical make-up [82]. In the U.S. 42,000 metric tons of oil are consumed annually as the active ingredient for insecticides alone [83]. Since the use of pesticides are limited in low-input systems, breeding for increased crop tolerance and resistance to economically devastating insect pests and pathogens could lead to improved yields, profit, and sustainability of these systems. Breeding for insect and disease tolerance can be more challenging than other limiting factors for several reasons. Many traits that improve tolerance to agricultural pests have been inadvertently breed away from in conventional breeding programs disregarding potential threats, however landraces in several crop species have been found to possess genotypes with improved resistance to pest and pathogen attack. Pest size, feeding strategy, reproduction, or infection behaviors can elicit different defense responses from the host plant, which can also differ among crop species. More importantly, insect and pathogenic pests are capable of adapting new behavioral and morphological responses to overcome the defenses of the targeted host [84]. However, several traits have been shown to promote crop tolerance and improve resistance to insect and pathogens for several economically valuable crop species.

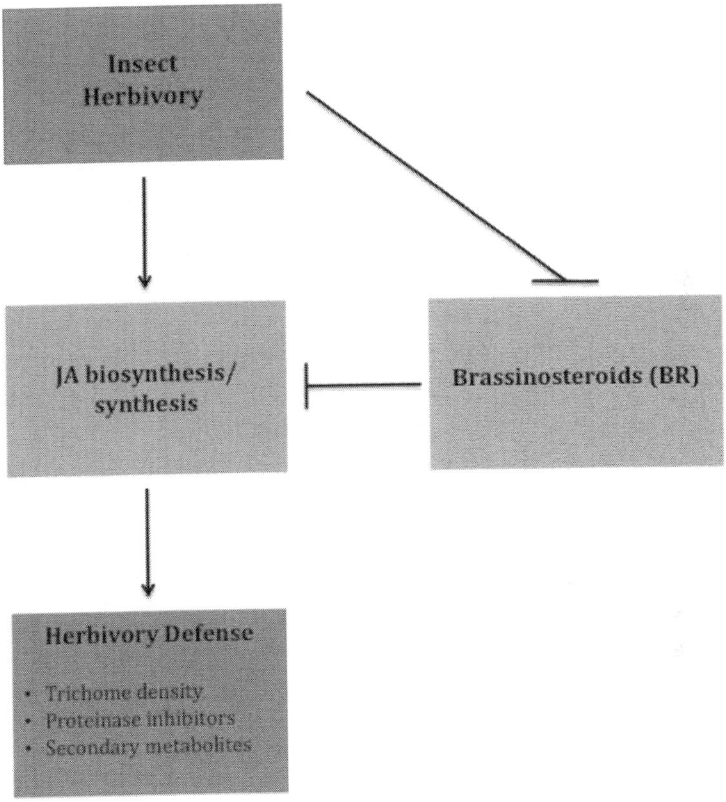

FIGURE 3: Model summarizing plant responses to insect herbivory related to hormonal pathways. The scheme highlights the negative crosstalk between the phytohormones brassinosteroid (BR) and jasmonic acid (JA), the latter being necessary in the deployment of defense mechanisms including trichome density, levels of proteinase inhibitors, and secondary metabolites [100,101].

Insect resistance and tolerance has been shown to be generally quantitative and polygenic [85]. Several traits were shown to specifically deter the herbivory of insect pests including changes in both epidermal and chemical composition, including leaf glossiness, cuticular wax, trichome density, and hormone production. Phenotypes that exhibit glossy leaves have shown increased resistance to insect feeding in several crop species, such as cabbage, soybean, common bean, maize, and sorghum [86-89]. Picoaga et al. [86] found that *Brassica* sp. with glossy leaf phenotypes were more resistant to insect herbivory. Eigenbrode and Espelie [87] suggested that reduced concentration and chemical composition of epicuticular lipids, common to glossy leaves, was the primary reason behind the increased crop resistance to insect infestation, ultimately effecting pest movement, feeding, and oviposition. Chemical composition of epicuticular lipids can be an important factor in deterring insect herbivory, crop devastation, and yield loss. It has been illustrated that aphid resistance is correlated to high concentrations of triacontarial (C_{30}) in alfalfa and β-amyrin in raspberries [87,90,91]. Several studies have implied that pesticides and other agricultural chemicals can affect the epicuticular lipid composition, reducing the ability of the crop to resist insect attack [92,93]. To date, multiple genes have been identified for the gloss leaf trait in cabbage, and 18 loci have been mapped using mutants of maize and sorghum [88,94,95]. Although leaf waxes can vary with crop age and be influenced by the production environment, they provide an avenue for breeders to develop varieties that are suited for specific agroecosystems while improving pest resistance, yield loss, and overall sustainability of low-input systems.

The most noted of changes to epidermal tissues leading to increased pest resistance is trichome density and morphology. Although not as effective against large or heavy insect pests, increased trichome density has been shown to alter the behavior of several small but economical agricultural pests. The defensive role of trichome density has been examined in several crop species. Diamondback moth resistance in *Arabidopsis*, green mite and mealybug resistance in cassava crops, and cabbage white butterfly larvae resistance in *Brassica* sp. were all associated with increased trichome density on both upper and lower leaf surfaces [84,96,97]. In addition, trichome morphology has also been shown to play an important role in limiting pest establishment within a crop. Sorghum genotypes with unicellular pointed trichomes were less susceptible to insect damage than genotypes possessing bicellular blunt trichomes [94]. Satish et al. [94] identified eight QTLs (quantitative trait loci) for trichome density using sorghum, two of which were specific for upper leaf surface, the remaining six specific to the lower leaf surface. Four QTL were identified in maize for trichome density by Lauter et al. [98], several of which were syntenic to those determined in sorghum. By breeding for increased trichome densities and beneficial morphology, improved resistance to insect herbivory, specific per host-herbivore relationship can be achieved.

Although several epidermal traits have been correlated to improved resistance to insect herbivory, recently much of the research has focused on molecular plant responses that improve tolerance to insect feeding. Several phytohormones are involved in a plant's defense response to insect herbivory, either directly or indirectly, however jasmonates (JA) appear to have the strongest involvement in response to insect feeding. Using hormone-related mutants of *Arabidopsis*, Abe et al. [99] found that JA played the most significant role in defense and tolerance to thrip feeding. Although JA plays an important role in anti-herbivory defense, it does not act alone. Brassinosteroids (BR) have been shown to have a significant, yet negative interaction with JA in the stimulation of herbivory defenses in tomato, specifically trichome development and regulation of proteinase inhibitors [100]. More recently, progress has been made to better understand BR and JA crosstalk involved with herbivore defense (Figure 3). Meldau et al. [101] determined that the SGT1 protein is involved in the accumulation of JA and when absent herbivory defense is reduced. In addition, BAK1, a co-receptor involved in BR signaling, also plays a role in JA accumulation, as well as involvement in altering levels of proteinase inhibitors [102]. By breeding plants with increased sensitivity to JA or insensitivity to BR, genotypes can be developed that will help improve insect resistance, reducing the need for heavy insecticide applications, while increasing the yield and sustainability of low-input systems.

The development of varieties resistant to common, crop specific pathogens of economic importance is essential to reducing the pesticides needed and the energy consumed by low-input systems to improve yield. Bacteria and fungi, like insects, are difficult to breed for based on the variety of way in which they infect crops and reproduce, as well as their ability to mutate in order to overcome the host's defense mechanisms. Pyramiding several resistance genes into one genotype is becoming a more widely used to develop durable resistance, with the hopes that the target pathogen will not undergo mutations that overcome all the resistant genes [103]. Several genotypes have been developed with improved resistance to economic pathogens via gene pyramiding including barley, rice, wheat, and tomato [104-107]. Liu et al. [107] developed durable and broad-spectrum powdery mildew resistance in wheat using several resistance genes, *Pm2, Pm4a,* and *Pm21*. By pyramiding the resistance genes *xa5, xa13, Xa21,* bacterial blight resistance was developed in rice [105]. However research has shown that to develop superior genotypes with durable resistance, alleles are necessary at more than one QTL [104,105]. Singh et al. [105] demonstrated that multiple members and combinations of resistant genes condition different responses to pathogen infection, with some genes and combinations being more effective at offering resistance. Pyramiding for disease tolerance will likely occur for specific host-pathogen interactions. By comparing Solanaceous crops (tomato, potato, pepper), Grube et al. [106] identified 12 cross-generic disease resistant genes (R genes). However, R genes with specificity to the same pathogen were only found twice at corresponding locations in different hosts. Improving the resistance of common agricultural pathogens via pyramiding will be laborious and require additional research, but will offer breeders an opportunity to develop genotypes with durable resistance, providing low-input producers with varieties less susceptibility to pathogen infection, less reliance on pesticide applications, while potentially improving yields.

11.4.5 PHOSPHORUS

Plants require three major mineral macronutrients (N-P-K) and a host of other essential micronutrients in order to develop properly. High-input agriculture relies heavily on fertilization with macronutrients, and fertilizer production industry supplies farmers with mostly inorganic macronutrients. Nitrogen (N) is captured from the atmosphere and reduced to ammonium using the Haber-Bosch process previously described. Potassium (K) is mined, with the current reserves expected to last for several centuries, thus not being of current concern. Phosphorus (P) is also mined, but its reserves are projected to only last between 50–130 years, given today's rate of application and the growing population's food demand as a guideline for production needs [1,108]. Phosphorus is not found as a free element on Earth, but instead is bound up as phosphates, typically found within inorganic rocks. Reserve supplies are even less evenly distributed than oil and are found primarily in China, U.S., Morocco, and in small South Pacific Islands [108]. In the U.S., phosphate is mined primarily from a single location in central Florida, supplying 75% of the phosphorus used by U.S. farmers, which corresponds to 25% of the world's phosphate reserves. However, the supply at this particular location is expected to only last several more decades [108]. Using the same modeling tools for analyzing oil production, phosphate follows a similar parabolic curve, with world production projected to reach its peak in 2030 [1]. Phosphorus is capable of leaching from sandy soil, and has been carelessly applied in over-abundance for decades, with a wasteful use of a finite resource as well as environmental pollution resulting in eutrophication of water bodies [1,108]. Three countries consume over 50% of the world phosphate annually mined, with China being the major consumer (30%) followed by India (15%), and the U.S. (11%) [109].

Phosphorus is considered a non-renewable resource, but there is possibility of it being recycled to some extent. During crop production, phosphorus is translocated from the soil to plant tissues, and subsequently consumed by humans and livestock. Little of the phosphorus available in the plant tissues is metabolically used by humans or livestock, and is therefore excreted [109]. Much of the phosphate in plant cells is stored as phytates (hexakisphosphate, IP6), which are not digested by monogastric animals. Phosphorus can then be recycled by collecting the excreted material, and reapplied to production fields in the form of manure or compost [1,108]. Even though recyclable, it is important to stress that once the supply of inorganic phosphorus has been exhausted, there is no other source nor is there any suitable phosphorus substitute in agriculture [1,108]. The consequence of phosphate depletion can be seen in Nauru, a South Pacific Island (in Micronesia, formerly known as Pleasant Island). After 90 years of intense phosphate mining (mostly consumed by the U.K., Australia, and New Zealand), 80% of central Nauru is now abandoned wasteland, and the supply is quickly approaching complete depletion, [108]. Production of phosphorus on the island nation has gone from 2.3 million tons (valued at $68/ton) in 1973 to a mere 250,000 tons (valued at only $44/ton) in 2001 [1]. The aggressive mining had adverse effects on vegetation and soil, destruction of the local ecology and economy, and depletion of the mineral content of the land itself [1,108].

According to the worldwide phosphorus production data, consumption and population growth are directly correlated [109]. Consumption and depletion of phosphate rocks are projected to rise with continued increases in population and food demand [109]. Breeding for varieties with higher phosphorus use efficiency could help to improve the worldwide management of this valuable resource, while providing enough food for the future. Factors to be considered in such breeding programs include improvement of root architecture, organic acid production and exudation, establishment of stronger mycorrhizal associations, more efficient phosphate uptake systems, and better phosphate physiology, which include less allocation of phosphate towards phytate biosynthesis and accumulation. Much of the research to improve phosphorus uptake efficiency has focused on improved morphology or physiology of the root system. In many soils the total amount of phosphorus can be high, however for the most part is present in organic forms that are unavailable to crops. One of the mechanisms used by crops produced in low-P systems is to alter root structure, allocating more carbon to the roots, increasing the root-to-shoot ratio [110,111]. *Arabidopsis* produced under phosphate deficiency shows modifications to the root architecture, redistributing energy from primary to lateral root growth [111], not to mention higher anthocyanin accumulation. Under P-deficient conditions, genotypes efficient in the acquisition of phosphorus have increased lateral root length allowing for greater exploration and foraging of the topsoil [112,113]. Change in root distribution has also been shown in tobacco, rape, spinach, and tomato [113-115]. Lynch and Brown [112] illustrated that genotypes with superior growth in low-P environments have root traits advantageous to topsoil foraging. It was also concluded that these inheritable traits are mediated by ethylene production and QTLs were identified through genetic mapping and used specifically for breeding towards improved phosphorus acquisition in low-input systems.

Most noted of the changes in root morphology due to phosphorus deficiency is improved root surface area, achieved by increases in length and density of root hairs. Narang et al. [116] showed that *Arabidopsis* developed long root hairs at high densities with high substrate penetration, ultimately improving the uptake of phosphorus per root length. Root hairs have been shown to be most effective at mining phosphorus from soil due to the large root surface area in direct contact with the soil and sustain high grain yields in low-P fields [117,118]. Demonstrated under controlled conditions, root hairs are the primary means of acquiring phosphorus from soil, contributing as much as 63% to the total phosphorus uptake [119]. In addition, Yan et al. [120] demonstrated a correlation between root hair length and phosphorus acquisition in field tests. Recent studies have shown that increased root hair development under low phosphorus conditions is under genetic control. Forty genes have been identified in *Arabidopsis* that are involved in root hair initiation and QTLs unique to low-P conditions [121,122]. By breeding varieties adapted to low phosphorus, possessing superior traits to acquire phosphorus could improve crop growth and potential yield.

Producing crops under phosphorus deficiency is difficult, but by increasing the density and length of root hairs, a crop's ability to acquire nutrients can significantly be improved. Also, more efficient membrane transport systems can be selected for to aid in efficient phosphorus uptake. In general, plants have two systems for phosphorus transport, a low-affinity (which operates at the millimolar scale) and a high-affinity system (which operates at the micromolar scale), the latter which has increased expression under low-P input [123]. Interaction between P deficiency and other factors (such as aluminum toxicity and micronutrient deficiency) should also be taken into consideration, but so far few studies have addressed this issue [124]. Several studies have reported phosphate transporters in multiple organs, including root, shoot, and reproductive tissues, but are found to have the greatest expression in root hairs [125,126]. A number of phosphorus transporter genes have been identified in various crops, and several *Arabidopsis* and tomato mutant genotypes possessing abnormal transporter expression have been described [123,127,128]. Although there has not been a clear correlation made between increased expression of high-affinity transporters and phosphorus acquisition, genetic variation among genotypes is well documented [126], indicating the potential to develop new crop varieties with increased potential to adapt to low-P availability.

Breeding for root systems that are more efficient at phosphorus acquisition could inadvertently lead to the development of varieties with improved mobilization of organic phosphorus reserves in the soil. Root apices exude a variety of organic acids, which can influence plant nutrition and provide an easily degradable nutrient source for soil microorganisms [129]. Of the organic acids exuded by roots under phosphorus deficiency, citrate, malate, and oxalate are the most effective at mobilizing soil phosphorus [130,131]. These organic acids can release unavailable phosphorus from bound minerals, allowing for the chelation of Al^{3+}, Fe^{3+}, and Ca^{2+} consequently freeing phosphorus and helping to alleviate P stress. Roots of white lupin growing under P stress exuded 20-40% more citrate and malate in comparison to roots provided sufficient supplies of phosphate [7,132]. Differences in the exudation of organic acids can be seen between crops under P-deficiency or not [133,134], suggesting potential to produce genotypes with improved ability to mobilize phosphate. Although the exact mechanism linking genetic regulation and the exudation of organic acids from root tips is largely unknown, gene expression data imply a complex coordinated induction of genes related to the synthesis, degradation, and utilization of citrate under P stress [123,135].

In addition to improving access of previously unavailable phosphate via rhizosphere acidification, exuded carboxylates promote microbial growth, and could potentially be used to exploit beneficial microbial relationships that might correlate with P bioavailability [129]. It has long been reported that beneficial relationships between crops and mycorrhizal fungi can improve availability and uptake of nutrients, in particular phosphorus [136]. Mycorrhizal fungi can increase phosphorus availability by exudating various organic acids themselves, freeing phosphates in the same manner as those exuded from plant roots. Colonization by beneficial fungi can lead to improved access of phosphorus by extending the crop's root system with mycorrhizal hyphae [137], indirectly increasing the root surface area for nutrient absorption and crop growth. Mycorrhizal hyphae work to improve nutrient acquisition by increasing their affinity for phosphorus ions and decreasing the concentration gradient required for more energy efficient absorption [138]. Benefits of mycorrhizal colonization have been observed mostly in organic and low-input systems with P deficiencies [79,139,140]. Studies have shown that maize produced under P deficient conditions has increased P acquisition and plant growth. However, this was not sustained as P concentrations were increased [141]. Additionally, biodiversity of AM fungi is greater in low-input production systems compared to high-input, likely due to the availability of nutrients making microbial symbiotic relationships obsolete and energy expensive to the crop [142]. Furthermore, Xavier and Germida [143] suggested that colonization by mycorrhizal fungi is correlated to yield responses in wheat, dependent on genotype and other advantageous root traits. The specific genetic mechanisms promoting symbioses between AM fungi and crop plants are not fully understood, although genotypic differences have been observed in maize, rice, and wheat [77,140,144]. Hetrick et al. [77] determined that landraces and traditional varieties developed prior to 1950 had greater reliance on mycorrhizal relationships than modern varieties. This implies that landraces and traditional varieties possess specific traits and genotypes that are beneficial in the development of symbiotic relationships with soil microbes. By reintroducing these favorable alleles into modern varieties, nutrient acquisition could improve, which may ultimately reduce the amount and need for phosphorus fertilizers.

The activity of certain enzymes may also prove to be valuable when selecting varieties for low P conditions. Acid phosphatases are ubiquitous enzymes present in various plant organs throughout development. They are responsible for providing phosphate to growing tissues during germination from stored phytate, remobilizing internal phosphate. Organic phosphate can breakdown reserves in the soil through exudation from roots into the rhizosphere when under low-input conditions [123,129]. Marschner et al. [145] found that P-efficient genotypes grown in P-depleted soils had greater phosphatase activity, which correlated to improved plant growth and nutrient uptake. Intracellular phosphatase activity when under P-stress, primarily functions to remobilize P from stored phytate and senescing tissues [123,146]. By breeding for increased phosphatase activity or any combination of the P-efficiency traits, crops produced in low-P soils will significantly improve the ability to acquire P whether from soil reserves or through remobilization of internally stored supplies, leading to varieties with optimal performance under stress conditions and increased sustainability of low-input agroecosystems. Unveiling these processes and associated genetic elements related to P physiology is extremely important and urgent in order to produce genotypes with enhanced root exudation and phosphate mobilization specifically in low-P conditions.

11.5 SCIENTIFIC AND NON-SYSTEMATIC ADVANCES: BREEDING STRATEGIES AND CONCEPTS

Low-input systems create a unique and complex environment, which are often composed of multiple factors limiting yield, making high yields difficult to achieve. However, low-input systems encompass a more sustainable agriculture due to improved management of on-farm resources. By increasing the availability of varieties that perform well under low-input conditions, the potential to meet the production demands of forthcoming populations could significantly be improved. By shifting breeding and selection methods toward low-input conditions and by making better use of local natural genetic variability, varieties that are best suited genetically and able to respond accordingly when exposed to stress conditions can improve the management of valuable, finite resources, as well as potentially decrease the energy used to produce sufficient quality food to people around the world. Several selection strategies could be successfully implemented to improve low-input breeding programs, such as participatory breeding [147-149]. It is important to notice, nonetheless, that the wide variety of cultural practices common to low-input production systems can create challenges for breeders.

Most new breeding strategies are built on the idea of natural selection. Allard and Hansche [150] originally stated that natural selection identifies superior crop genotypes, which make up a greater portion of the population in over time. After many generations, natural selection produces genotypes well suited to unpredictable and stressful environments that are common to low-input systems [151]. By breeding under high-input conditions, the opportunity to exploit advantageous genetic differences at low input levels is lost, resulting in exclusion of important alleles needed to provide adequate and superior varieties [38]. Ceccarelli [38] suggested the need for local breeding programs with active participation from farmers to achieve increased sustainability in low-input agricultural systems. The idea being that local varieties or those most successfully produced in stress environments will possess traits or adaptations that are advantageous to crop growth, yield, and consumer expectations. Building on the idea of natural selection, landraces, creole, and heirloom varieties [152], are best suited to the environment in which they originated, having adapted according to the selection pressures provided by the local agroecosystem. Incorporation of valuable traits from heirlooms and landraces with high yielding varieties can help optimize production in low-input systems, and thus the potential of these systems to fulfill the food demand of the future.

At this point, it is important to briefly distinguish breeding methods that are used for autogamous (self-pollinating) species from those employed for allogamous (outcrossing) crops. Autogamous crops (e.g., rice, wheat, barley, oat, common bean, soybean, lentil, tomato) tolerate inbreeding, thus allowing sexual propagation of highly homozygous varieties, meaning that they can produce offspring that is genetically identical to the parental line when a high level of inbreeding is achieved. Thus, breeding programs established for these crops rely on creating genetic combinations by artificially crossing genotypes with traits of interest and undergoing further rounds of selection and self-pollination to reach variety stability, in which the new variety present the trait(s) of interest. Allogamous species (e.g., maize, rye, pearl millet, cotton, sugar beet, canola, squash, cucumber, papaya, cassava), however, show poorer genetic performance under full homozygosity (i.e., inbred lines), while benefiting of the heterozygous state via outcrossing. For these species, highly inbred parental lines are usually selected and hybrid seeds are produced by combining parental lines derived from distinct populations, in order to ensure heterozygosity in first generation (F_1) seeds. The performance of F_2 populations derived from hybrid seeds tends to lose agronomical performance in relation to its parental (F_1) population, probably due to a higher degree of homozygosity. Thus, in commercial settings, the maintenance of high yields of allogamous hybrid crops depends largely on the purchase of expensive seeds year after year, encompassing a high-input supply.

Thus, breeding for autogamous species is more straightforward than for allogamous crops regarding selection and maintenance of the genetic identity in the final bred variety via seed propagation. Likewise, allogamous species with a long juvenile phase (such as many fruit and nut trees) or difficult seed production (e.g., banana, sugarcane, garlic) are conventionally crossed to produce a segregating generation, and selection is carried out already in the F_1 generation, while the genetic identity of the (hybrid) cultivar is guaranteed thereafter via clonal (vegetative) propagation. In contrast, for annual allogamous crops, production in a low-input system must rely on the performance of the whole population instead of focusing on yields of single individuals, since the key for performance maintenance over generations depends on a diversified genetic pool within the population. Selection under low input in allogamous species will thus increase the frequency of alleles in the population that are responsible for acceptable yields at marginal conditions, rather than producing single, genetically homogenous hybrid lines that may lack adaptive alleles due to selection at high-input conditions. This strategy has been successfully demonstrated for maize [153]. Evidence of feasibility of breeding for low input systems by using landraces as starting material has been recently shown for barley [154]. Outperformance of selection under low-input over high-input systems has been demonstrated by Ceccarelli et al. [155], in which barley cultivars selected under low-input yielded up to 54% more under stress conditions than the cultivars selected under high-input, in the same conditions.

Low heritability of agronomic traits (thus, little potential for crop improvement) in local populations has been reported for oat [156] and faba bean [157]. In these cases, the genetic basis of the population under study might be too narrow and alleles of interest may have been lost or not locally introduced. These cases are more common in regions far from the centers of speciation or distribution of the crop and it can often be resolved with introduction of new germplasm in the population. This underscores how important breeding techniques developed over the last century should be applied to advance local breeding initiatives, in order to create programs to attend the specific demands and needs of farmers in a bottom-up approach, but without forfeiting the scientific approach. Diverse breeding techniques have been developed to take advantage of natural genetic variation available and to actively involve the participation of farmers in the process.

Evolutionary breeding (EB) refers to a technique in which mass selection is used, favored by natural selection, concentrating largely on high-yield genotypes [13]. Participatory plant breeding (PPB) refers to selection methods which were developed in response to meet the needs of low-input producers who were largely without suitable varieties for adverse conditions, more recently this technique has also benefited organic producers [13]. This method of selection, which occurs on-site, resulting in 'island effects' with significant, but specific, local adaptations that improve crop stability, farm sustainability, and increase local marketing opportunities [12,152].

Evolutionary participatory breeding (EPB) is a marriage of both techniques aforementioned, driven by natural selection on genetically diverse populations in order to utilize genetic variability, and consequently the ability to adapt to unpredictable stresses, with active site-specific, on-farm variety selection. Both participatory breeding programs mentioned emphasize communication between breeder and producers with the selection of genotypes including the knowledge and expertise of local farmers. The goal of these programs is to produce varieties that are well suited to the environment and production practices, ultimately increasing the sustainability and profitability of low-input systems. Participatory breeding strategies have been successful in developing improved varieties that are able to adapt to the low-input production environment, and are more locally accepted than modern varieties. Trocuhe et al. [29] found that producers can consistently select varieties that are best suited to the production environment, adding that inclusion of local producers to breeding programs will greatly benefit food production in limiting systems. Initial participatory breeding programs have successfully led to the development of varieties of a global significance including barley [26], sorghum [29,34], maize [158], and wheat [153]. Several new programs have also been initiated and have begun to include locally valued crops such as common bean, cassava, and potato [159].

It is fact that improving yields in low-input systems will likely not result in improvement in the availability or use of just one limiting resource, but will require concomitant improvements of several production limitations. This may require advanced breeding techniques, from genetics and statistics stand points, to achieve superior genotypes and varieties that are more sustainable in low-input production. Participatory breeding programs have traditionally relied on low-cost, low-technology techniques, due to the lack of resources and the involvement of the scientific community, at large. However, the understanding of their potential for sustainable agriculture and global resource savings should foster more funding of these programs to allow the use of more effective breeding approaches. The use of molecular markers has been implicated as a quite possible avenue to breed varieties with superior genotypes, when phenotypic selection becomes unreliable [65]. Marker assisted selection (MAS) uses genetic markers or specific genetic sequences that have been determined to be associated with known locus linked to a desired trait. This technique is costly at first as favorable genomic regions (QTLs) are identified, but it saves time thereafter. After QTLs have been detected, closely linked markers can be used to trace, transfer and accumulate (trait pyramidation) the valued genetic regions into one superior genotype at a much faster speed than conventional breeding [65]. This particular method is also advantageous in low-input breeding because varieties can be further developed without recurrent field evaluations and during off-season, since it is based on linkage of the trait of interest with genomic regions revealed by the markers that are correlated with better performance, allowing for the quicker development of varieties. It is possible that current varieties, landraces, or heirlooms will serve as starting material or genotype, and that molecular markers will be used to transfer desired genetic segments from other genotypes in order to develop elite varieties specifically aiming for low-input production systems.

Although MAS has been key in developing modern varieties, it has primarily been successful in manipulating a few traits controlled by major effect genes [160]. Unfortunately this method has been insufficient when improving polygenic or quantitative traits that are controlled by several small effect genes [160]. A relatively new method, genome-wide or genomic selection has been developed to overcome the limitations of MAS. Genomic selection is a form of MAS that calculates breeding values by simultaneously analyzing all markers and phenotyping across an entire genome [161]. These scores can then be applied to model parameters used to estimate the value of future breeding lines with only marker data [160]. In addition this technique can be used without prior knowledge of marker trait associations and allowing for selection of multiple QTLs linked to small effect genes [161]. An important distinction to note is that breeding lines developed using genomic selection are not primarily evaluated on phenotypic response but on genomic information shared across other breeding lines, locations, and growth conditions, resulting in the development of varieties with increased stability and the increased ability to adapt to low yielding conditions [160]. Advances in several crops have been made using this method, including wheat, maize, and barley, and while progress is slow, this technique increases the potential for development of varieties that are specific for low-input production systems.

11.6 CONCLUSIONS

Overall, improvement in agricultural sustainability by means of increasing yields of low-input production systems is not only possible, but also urgently needed. By using breeding methods that are geared to the common limitations experienced by farmers around the globe, varieties with superior traits and adaptations can be achieved. Increasing the availability of superior varieties specifically bred to low-input systems, either through traditional or advanced breeding methods will improve agricultural sustainability and global resource management, as well as decrease the energy demanded for food production during a time of historic global relevance as population peaks and valuable finite resources decline.

The potential impact of using breeding for low-input conditions for a more sustainable agriculture is great, and indeed its feasibility has been demonstrated for many crops, both autogamous and allogamous. However, the use of local crop breeding initiatives for low-input systems requires mobilization of most immediate stakeholders, who unfortunately are often demobilized, decapitalized small farmers and peasants. Government actions worldwide and throughout history have largely neglected this group. Nonetheless, it is imperative and urgent that now, as world resources are becoming scarce, not only small farmers but also commercial agriculture embrace a more rational use of resources to produce enough food and raw materials for all. Government intervention will certainly be required to allow small farmers to continue cultivating the land, whereas also commercial farmers will need to face a paradigm shift towards sustainability to guarantee the future of the next generation in a superpopulated world.

REFERENCES

1. Cordell, D.; Drangert, J.; White, S. The story of phosphorus: Global food security and food for thought. Glob. Environ. Change 2009, 19, 292-305.
2. UNPD. World Population Prospects, the 2010 Revision; United Nations Population Division (UN DESA): New York, NY, USA, 2011.
3. Aleklett, K.; Hook, M.; Jakobsson, K.; Lardelli, M.; Snowden, S.; Soderbergh, B. The peak of the oil age—Analyzing the world oil production reference scenario in world energy outlook 2008. Energy Policy 2010, 38, 1398-1414.
4. Maggio, G.; Cacciola, G. A variant of the Hubbert curve for world oil production forecasts. Energy Policy 2009, 37, 4761-4770.
5. FAO. 2050: A Third More Mouths to Feed. Food and Agriculture Organization: Rome, Italy, 2009. Available online: http://www.fao.org/news/story/en/item/35571/ (accessed on 21 September 2011).
6. De Fraiture, C.; Wichelns, D. Satisfying future water demands for agriculture. Agric. Water Management 2010, 97, 502-511.
7. Vance, C.P. Symbiotic nitrogen fixation and phosphorus acquisition. Plant nutrition in a world of declining renewable resources. Plant Physiol. 2001, 127, 390-397.
8. Smil, V. Nitrogen in crop production: An account of global flows. Glob. Biogeochem. Cycles 1999, 13, 647-662.
9. FAO. Fertilizer Archive; Food and Agriculture Organization: Rome, Italy, 2011.
10. Tilman, D.; Fargione, J.; Wolff, B.; D'Antonio, C.; Dobson, A.; Howarth, R.; Schindler, D.; Schlesinger, W.H.; Simberloff, D.; Swackhamer, D. Forecasting agriculturally driven global environmental change. Science 2001, 292, 281-284.
11. IFA. World Fertilizer Consumption; International Fertilizer Industry Association: Paris, France, 2011. Available online: http://www.fertilizer.org/ifa/ifadata/search (accessed on 21 September 2011).
12. Phillips, S.L.; Wolfe, M.S. Evolutionary plant breeding for low input systems. J. Agric. Sci. 2005, 143, 245-254.
13. Murphy, K.; Lammer, D.; Lyon, S.; Carter, B.; Jones, S.S. Breeding for organic and low-input farming systems: An evolutionary-participatory breeding method for inbred cereal grains. Renew. Agric. Food Syst. 2005, 20, 48-55.
14. Lee, R. The outlook for population growth. Science 2011, 333, 569-573.
15. UNPD. World Population Prospects, the 2008 Revision; United Nations Population Division (UN DESA): New York, NY, USA, 2009.
16. Roberts, L. 9 Billion? Science 2011, 333, 540-543.
17. FAO. World Agriculture Towards 2030/2050; Food and Agriculture Organization: Rome, Italy, 2006.
18. FAO/IISA. Global Agro-Ecological Zones; Food and Agriculture Organization / IISA: Rome, Italy and London, UK, 2000.

19. FAO Statistics Division. ResouceSTAT. 2008; Food and Agriculture Organization: Rome, Italy, 2011.
20. Beddington, J. Food Security: Contributions from science to a new and greener revolution. Philos. Trans. R. Soc. B 2010, 365, 61-71.
21. Fischer, G.; Shah, M. Farmland Investments and Food Security, Statistical Annex Report Prepared Under World-Bank-IIASA; International Institute for Applied Systems Analysis: Laxenburg, Austria, 2010.
22. Lambin, E.F.; Meyfroidt, P. Global land use change, economic globalization, and the looming land scarcity. Proc. Natl. Acad. Sci. U. S. A. 2011, 108, 3465-3472.
23. Alig, R.J.; Kline, J.D.; Lichtenstein, M. Urbanization on the US landscape: Looking ahead in the 21st century. Landsc. Urban Plann. 2004, 69, 219-234.
24. Harvey, M.; Pilgrim, S. The new competition for land: Food, energy, and climate change. Food Policy 2011, 36, S40-S51.
25. Sthapit, B.; Rana, R.; Eyzaguirre, P.; Jarvis, D. The value of plant genetic diversity to resource-poor farmers in nepal and vietnam. Int. J. Agric. Sustain. 2008, 6, 148-166.
26. Abay, F.; Bjornstad, A. Specific adaptation of barley varieties in different locations in Ethiopia. Euphytica 2009, 167, 181-195.
27. Ceccarelli, S. Specific adaptation and breeding for marginal conditions. Euphytica 1994, 77, 205-219.
28. Toure, A.; Becker, M.; Johnson, D.E.; Kone, B.; Kossou, D.K.; Kiepe, P. Response of lowland rice to agronomic management under different hydrological regimes in an inland valley of Ivory Coast. Field Crops Res. 2009, 114, 304-310.
29. Trouche, G.; Aguirre Acuna, S.; Castro Briones, B.; Gutierrez Palacios, N.; Lancon, J. Comparing decentralized participatory breeding with on-station conventional sorghum breeding in Nicaragua: I. Agronomic performance. Field Crops Res. 2011, 121, 19-28.
30. Guarda, G.; Padovan, S.; Delogu, G. Grain yield, nitrogen-use efficiency and baking quality of old and modern Italian bread-wheat cultivars grown at different nitrogen levels. Eur. J. Agron. 2004, 21, 181-192.
31. Ortiz, R.; Braun, H.; Crossa, J.; Crouch, J.H.; Davenport, G.; Dixon, J.; Dreisigacker, S.; Duveiller, E.; He, Z.; Huerta, J.; et al. Wheat genetic resources enhancement by the international maize and wheat improvement center (CIMMYT). Genet. Resour. Crop Evol. 2008, 55, 1095-1140.
32. Reynolds, M.P.; Borlaug, N.E. Impacts of breeding on international collaborative wheat improvement. J. Agric. Sci. 2006, 144, 3-17.
33. Mandal, N.P.; Sinha, P.K.; Variar, M.; Shukla, V.D.; Perraju, P.; Mehta, A.; Pathak, A.R.; Dwivedi, J.L.; Rathi, S.P.S.; Bhandarkar, S.; et al. Implications of genotype × input interactions in breeding superior genotypes for favorable and unfavorable rainfed upland environments. Field Crops Res. 2010, 118, 135-144.
34. Vom Brocke, K.; Trouche, G.; Weltzien, E.; Barro-Kondombo, C.P.; Goze, E.; Chantereau, J. Participatory variety development for sorghum in Burkina Faso: Farmers' selection and farmers' criteria. Field Crops Res. 2010, 119, 183-194.
35. Yapi, A.M.; Kergna, A.O.; Debrah, S.K.; Sidibe, A.; Sanogo, O. Analysis of the Economic Impact of Sorghum and Millet Research in Mali; International Crops Research Institute for the Semi-Arid Tropics: Andhra Pradesh, India, 2000.

36. Murphy, K.M.; Campbell, K.G.; Lyon, S.R.; Jones, S.S. Evidence of varietal adaptation to organic farming systems. Field Crops Res. 2007, 102, 172-177.
37. Zeven, A.C. Landraces: A review of definitions and classifications. Euphytica 1998, 104, 127-139.
38. Ceccarelli, S. Adaptation to low high input cultivation. Euphytica 1996, 92, 203-214.
39. Seckler, D.; Amarasinghe, U.; Molden, D.; de Silva, R.; Barker, R. World Water Demand and Supply, 1990 to 2025: Scenarios and Issue; International Water Management Institute (IWMI): Colombo, Sri Lanka, 1998.
40. Markwei, C.; Ndlovu, L.; Robinson, E.; Shah, W. International Assessment of Agriculture Knowledge, Science, and Technology for Development (IAASTD) - Sub Saharan Africa Summary for Decision Makers. Available online: http://www.agassessment.org/docs/ SSA_SDM_220408_Final.pdf (accessed on 21 September 2011).
41. Richards, R.A.; Rebetzke, G.J.; Condon, A.G.; van Herwaarden, A.F. Breeding opportunities for increasing the efficiency of water use and crop yield in temperate cereals. Crop Sci. 2002, 42, 111-121.
42. Blum, A. Drought Resistance, Water-use efficiency, and yield potential—Are they compatible, dissonant, or mutually exclusive? Aust. J. Agric. Res. 2005, 56, 1159-1168.
43. Thompson, A.J.; Andrews, J.; Mulholland, B.J.; McKee, J.M.T.; Hilton, H.W.; Horridge, J.S.; Farquhar, G.D.; Smeeton, R.C.; Smillie, I.R.A.; Black, C.R.; et al. Overproduction of abscisic acid in tomato increases transpiration efficiency and root hydraulic conductivity and influences leaf expansion. Plant Physiol. 2007, 143, 1905-1917.
44. Condon, A.G.; Richards, R.A.; Rebetzke, G.J.; Farquhar, G.D. Improving intrinsic water-use efficiency and crop yield. Crop Sci. 2002, 42, 122-131.
45. Condon, A.G.; Richards, R.A.; Rebetzke, G.J.; Farquhar, G.D. Breeding for high water-use efficiency. J. Exp. Bot. 2004, 55, 2447-2460.
46. Iuchi, S.; Kobayashi, M.; Taji, T.; Naramoto, M.; Seki, M.; Kato, T.; Tabata, S.; Kakubari, Y.; Yamaguchi-Shinozaki, K.; Shinozaki, K. Regulation of drought tolerance by gene manipulation of 9-cis-epoxycarotenoid dioxygenase, a key enzyme in abscisic acid biosynthesis in Arabidopsis. Plant J. 2001, 27, 325-333.
47. Thompson, A.J.; Jackson, A.C.; Symonds, R.C.; Mulholland, B.J.; Dadswell, A.R.; Blake, P.S.; Burbidge, A.; Taylor, I.B. Ectopic expression of a tomato 9-cis-epoxycarotenoid dioxygenase gene causes over-production of abscisic acid. Plant J. 2000, 23, 363-374.
48. Qin, X.; Zeevaart, J. Overexpression of a 9-cis-epoxycarotenoid dioxygenase gene in Nicotiana plumbaginifolia increases abscisic acid and phaseic acid levels and enhances drought tolerance. Plant Physiol. 2002, 128, 544-551.
49. Ali, M.; Jensen, C.R.; Mogensen, V.O.; Andersen, M.N.; Henson, I.E. Root signalling and osmotic adjustment during intermittent soil drying sustain grain yield of field grown wheat. Field Crops Res. 1999, 62, 35-52.
50. Deyholos, M.K. Making the most of drought and salinity transcriptomics. Plant Cell Environ. 2010, 33, 648-654.
51. Karaba, A.; Dixit, S.; Greco, R.; Aharoni, A.; Trijatmiko, K.R.; Marsch-Martinez, N.; Krishnan, A.; Nataraja, K.N.; Udayakumar, M.; Pereira, A. Improvement of wa-

ter use efficiency in rice by expression of HARDY, an Arabidopsis drought and salt tolerance gene. Proc. Natl. Acad. Sci. USA 2007, 104, 15270-15275.
52. Bouchabke-Coussa, O.; Quashie, M.; Seoane-Redondo, J.; Fortabat, M.; Gery, C.; Yu, A.; Linderme, D.; Trouverie, J.; Granier, F.; Teoule, E.; et al. ESKIMO1 is a key gene involved in water economy as well as cold acclimation and salt tolerance. BMC Plant Biol. 2008, 8, 125-151.
53. Lefebvre, V.; Fortabat, M.; Ducamp, A.; North, H.M.; Maia-Grondard, A.; Trouverie, J.; Boursiac, Y.; Mouille, G.; Durand-Tardif, M. ESKIMO1 disruption in Arabidopsis alters vascular tissue and impairs water transport. PLoS One 2011, 6, e16645.
54. Reynolds, M.; Tuberosa, R. Translational research impacting on crop productivity in drought-prone environments. Curr. Opin. Plant Biol. 2008, 11, 171-179.
55. Franks, S.J. Plasticity and evolution in drought avoidance and escape in the annual plant Brassica rapa. New Phytol. 2011, 190, 249-257.
56. Bouchabke, O.; Chang, F.; Simon, M.; Voisin, R.; Pelletier, G.; Durand-Tardif, M. Natural variation in Arabidopsis thaliana as a tool for highlighting differential drought responses. PLoS One 2008, 3, e1705.
57. Duncan, R.C.; Youngquist, W. Encircling the peak of world oil production. Natl. Resour. Res. 1999, 8, 219-232.
58. Baligar, V.; Fageria, N.; He, Z. Nutrient use efficiency in plants. Commun. Soil Sci. Plant Anal. 2001, 32, 921-950.
59. USDA. Consumption of Plant Nutrients; United States Department of Agriculture: Washington, DC, USA, 2011.
60. Peterson, T.A.; Russelle, M.P. Alfalfa and the nitrogen-cycle in the corn belt. J. Soil Water Conserv. 1991, 46, 229-235.
61. Pfieffer, D.A. Eating Fossil Fuels: Oil, Food and the Coming Crisis in Agriculture; New Society Publishers: Gabriola, Canada, 2006.
62. Dawson, J.C.; Huggins, D.R.; Jones, S.S. Characterizing nitrogen use efficiency in natural and agricultural ecosystems to improve the performance of cereal crops in low-input and organic agricultural systems. Field Crops Res. 2008, 107, 89-101.
63. Bertin, P.; Gallais, A. Genetic variation for nitrogen use efficiency in a set of recombinant maize inbred lines I. Agrophysiological results. Maydica 2000, 45, 53-66.
64. Gallais, A.; Hirel, B. An approach to the genetics of nitrogen use efficiency in maize. J. Exp. Bot. 2004, 55, 295-306.
65. Gallais, A.; Coque, M. Genetic variation and selection for nitrogen use efficiency in maize: A synthesis. Maydica 2005, 50, 531-547.
66. Basra, A.S.; Goyal, S.S. Mechanisms of Improved Nitrogen-use Efficiency in Cereals. In Quantitative Genetics, Genomics and Plant Breeding; Kang, M.S., Ed.; CABI Publishing: Oxfordshire, UK, 2002; p. 288.
67. Hirel, B.; Bertin, P.; Quillere, I.; Bourdoncle, W.; Attagnant, C.; Dellay, C.; Gouy, A.; Cadiou, S.; Retailliau, C.; Falque, M.; et al. Towards a better understanding of the genetic and physiological basis for nitrogen use efficiency in maize. Plant Physiol. 2001, 125, 1258-1270.
68. Duncan, R.R.; Baligar, V.C. Genetics Breeding and Physiological Mechanisms of Nutrient Uptake and use Efficiency an Overview. In Crop as Enhancers of Nutrient Use; Academic Press: San Deigo, CA, USA, 1990; p. 36.

69. Jordi, W.; Schapendonk, A.; Davelaar, E.; Stoopen, G.M.; Pot, C.S.; de Visser, R.; van Rhijn, J.A.; Gan, S.; Amasino, R.M. Increased cytokinin levels in transgenic PSAG12-IPT tobacco plants have large direct and indirect effects on leaf senescence, photosynthesis and N partitioning. Plant Cell Environ. 2000, 23, 279-289.
70. Wingler, A.; Purdy, S.; MacLean, J.A.; Pourtau, N. The role of sugars in integrating environmental signals during the regulation of leaf senescence. J. Exp. Bot. 2006, 57, 391-399.
71. Presterl, T.; Seitz, G.; Landbeck, M.; Thiemt, E.M.; Schmidt, W.; Geiger, H.H. Improving nitrogen-use efficiency in European maize: estimation of quantitative genetic parameters. Crop Sci. 2003, 43, 1259-1265.
72. Zaidi, P.H.; Srinivasan, G.; Sanchez, C. Relationship between line Per Se and cross performance under low nitrogen fertility in tropical maize (Zea mays L.). Maydica 2003, 48, 221-231.
73. Spano, G.; di Fonzo, N.; Perrotta, C.; Platani, C.; Ronga, G.; Lawlor, D.W.; Napier, J.A.; Shewry, P.R. Physiological characterization of 'stay green' mutants in durum wheat. J. Exp. Bot. 2003, 54, 1415-1420.
74. Sanginga, N.; Lyasse, O.; Singh, B.B. Phosphorus use efficiency and nitrogen balance of cowpea breeding lines in a low P soil of the derived savanna zone in West Africa. Plant Soil 2000, 220, 119-128.
75. Azcon, R.; Ruiz-Lozano, J.; Rodriguez, R. Differential contribution of arbuscular mycorrhizal fungi to plant nitrate uptake (N-15) under increasing n supply to the soil. Can. J. Botany-Revue Can. De Bot. 2001, 79, 1175-1180.
76. Hetrick, B.A.D.; Wilson, G.W.T.; Cox, T.S. Mycorrhizal dependence of modern wheat cultivars and ancestors—A synthesis. Can. J. Botany-Revue Can. De Bot. 1993, 71, 512-518.
77. Hetrick, B.A.D.; Wilson, G.W.T.; Cox, T.S. Mycorrhizal dependence of modern wheat-varieties, landraces, and ancestors. Can. J. Botany-Revue Can. De Bot. 1992, 70, 2032-2040.
78. Jeffries, P.; Gianinazzi, S.; Perotto, S.; Turnau, K.; Barea, J.M. The contribution of arbuscular mycorrhizal fungi in sustainable maintenance of plant health and soil fertility. Biol. Fertil. Soils 2003, 37, 1-16.
79. Mercy, M.A.; Shivashankar, G.; Bagyaraj, D.J. Mycorrhizal colonization in cowpea is host-dependent and heritable. Plant Soil 1990, 121, 292-294.
80. Galvan, G.A.; Paradi, I.; Burger, K.; Baar, J.; Kuyper, T.W.; Scholten, O.E.; Kik, C. Molecular diversity of arbuscular mycorrhizal fungi in onion roots from organic and conventional farming systems in the Netherlands. Mycorrhiza 2009, 19, 317-328.
81. Sanders, I.R.; Croll, D. Arbuscular mycorrhiza: The challenge to understand the genetics of the fungal partner. Annu. Rev. Genet. 2010, 44, 271-292.
82. Helsel, Z.R. Energy in Pesticide Production and Use. In Encyclopedia of Pest Management; Pimental, D., Ed.; Elsevier: New York, NY, USA, 2002; p. 177.
83. Gianessi, L.; Reigner, N. Pesticide Use in US Crop Production: 2002; Crop Protection Research Institute: Washington, DC, USA, 2006.
84. Hanley, M.E.; Lamont, B.B.; Fairbanks, M.M.; Rafferty, C.M. Plant structural traits and their role in anti-herbivore defence. Perspect. Plant Ecol. Evol. Syst. 2007, 8, 157-178.

85. Ojwang, P.P.O.; Melis, R.; Githiri, M.; Songa, J.M. Breeding options for improving common bean for resistance against bean fly (Ophiomyia spp.): A review of research in Eastern and Southern Africa. Euphytica 2011, 179, 363-371.
86. Picoaga, A.; Cartea, M.E.; Soengas, P.; Monetti, L.; Ordas, A. Resistance of kale populations to lepidopterous pests in Northwestern Spain. J. Econ. Entomol. 2003, 96, 143-147.
87. Eigenbrode, S.D.; Espelie, K.E. Effects of plant epicuticular lipids on insect herbivores. Annu. Rev. Entomol. 1995, 40, 171-194.
88. Beattie, G.A.; Marcell, L.M. Effect of alterations in cuticular wax biosynthesis on the physicochemical properties and topography of maize leaf surfaces. Plant Cell Environ. 2002, 25, 1-16.
89. Talekar, N.S.; Shelton, A.M. Biology, ecology, and management of the diamondback moth. Annu. Rev. Entomol. 1993, 38, 275-301.
90. Robertson, G.W.; Griffiths, D.W.; Birch, A.N.E.; Jones, A.T.; McNicol, J.W.; Hall, J.E. Further evidence that resistance in raspberry to the virus vector aphid, Amphorophora idaei, is related to the chemical-composition of the leaf surface. Ann. Appl. Biol. 1991, 119, 443-449.
91. Bergman, D.K.; Dillwith, J.W.; Zarrabi, A.A.; Caddel, J.L.; Berberet, R.C. Epicuticular lipids of alfalfa relative to its susceptibility to spotted alfalfa aphids (Homoptera, Aphididae). Environ. Entomol. 1991, 20, 781-785.
92. Cabras, P.; Angioni, A.; Garau, V.L.; Melis, M.; Pirisi, F.M.; Minelli, E.V. Effect of epicuticular waxes of fruits on the photodegradation of fenthion. J. Agric. Food Chem. 1997, 45, 3681-3683.
93. Tamura, H.; Knoche, M.; Bukovac, M.J. Evidence for surfactant solubilization of plant epicuticular wax. J. Agric. Food Chem. 2001, 49, 1809-1816.
94. Satish, K.; Srinivas, G.; Madhusudhana, R.; Padmaja, P.G.; Reddy, R.N.; Mohan, S.M.; Seetharama, N. Identification of quantitative trait loci for resistance to shoot fly in Sorghum sorghum bicolor (L.) Moench. Theor. Appl. Genet. 2009, 119, 1425-1439.
95. Brooks, T.D.; Willcox, M.C.; Williams, W.P.; Buckley, P.M. Quantitative trait loci conferring resistance to fall armyworm and southwestern corn borer leaf feeding damage. Crop Sci. 2005, 45, 2430-2434.
96. Handley, R.; Ekbom, B.; Agren, J. Variation in trichome density and resistance against a specialist insect herbivore in natural populations of Arabidopsis thaliana. Ecol. Entomol. 2005, 30, 284-292.
97. Mahungu, N.M.; Dixon, A.G.O.; Kumbira, J.M. Breeding cassava for multiple pest resistance in Africa. African Crop Sci. J. 1994, 2, 539-552.
98. Lauter, N.; Gustus, C.; Westerbergh, A.; Doebley, J. The inheritance and evolution of leaf pigmentation and pubescence in Teosinte. Genetics 2004, 167, 1949-1959.
99. Abe, H.; Ohnishi, J.; Narusaka, M.; Seo, S.; Narusaka, Y.; Tsuda, S.; Kobayashi, M. Function of jasmonate in response and tolerance of Arabidopsis to thrip feeding. Plant Cell Physiol. 2008, 49, 68-80.
100. Campos, M.L.; de Almeida, M.; Rossi, M.L.; Martinelli, A.P.; Litholdo Junior, C.G.; Figueira, A.; Rampelotti-Ferreira, F.T.; Vendramim, J.D.; Benedito, V.A.; Pereira Peres, L.E. Brassinosteroids interact negatively with jasmonates in the formation of anti-herbivory traits in tomato. J. Exp. Bot. 2009, 60, 4346-4360.

101. Meldau, S.; Baldwin, I.T.; Wu, J. SGT1 regulates wounding- and herbivory-induced jasmonic acid accumulation and nicotiana attenuata's resistance to the specialist lepidopteran herbivore Manduca sexta. New Phytol. 2011, 189, 1143-1156.
102. Yang, D.; Hettenhausen, C.; Baldwin, I.T.; Wu, J. BAK1 regulates the accumulation of jasmonic acid and the levels of trypsin proteinase inhibitors in Nicotiana attenuata's responses to herbivory. J. Exp. Bot. 2011, 62, 641-652.
103. McDonald, B.A.; Linde, C. Pathogen population genetics, evolutionary potential, and durable resistance. Annu. Rev. Phytopathol. 2002, 40, 349-379.
104. Castro, A.J.; Chen, X.M.; Hayes, P.M.; Johnston, M. Pyramiding Quantitative trait locus (QTL) alleles determining resistance to barley stripe rust: Effects on resistance at the seedling stage. Crop Sci. 2003, 43, 651-659.
105. Singh, S.; Sidhu, J.S.; Huang, N.; Vikal, Y.; Li, Z.; Brar, D.S.; Dhaliwal, H.S.; Khush, G.S. Pyramiding three bacterial blight resistance genes (xa5, xa13 and xa21) using marker-assisted selection into indica rice cultivar PR106. Theor. Appl. Genet. 2001, 102, 1011-1015.
106. Grube, R.C.; Radwanski, E.R.; Jahn, M. Comparative genetics of disease resistance within the Solanaceae. Genetics 2000, 155, 873-887.
107. Liu, J.; Liu, D.; Tao, W.; Li, W.; Wang, S.; Chen, P.; Cheng, S.; Gao, D. Molecular marker-facilitated pyramiding of different genes for powdery mildew resistance in wheat. Plant Breed. 2000, 119, 21-24.
108. Déry, P.; Anderson, B. Peak Phosphorus. Energy Bull. 2007. Published online: 13 August 2007. http://www.energybulletin.net/node/33164.
109. Smit, A.L.; Bindraban, P.S.; Schröder, J.J.; Conijn, J.G.; van der Meer, H.G. Phosphorus in Agriculture Global Resources, Trends, and Development; Plant Research International: Wageningen, The Netherlands, 2009.
110. Nielsen, K.L.; Eshel, A.; Lynch, J.P. The effect of phosphorus availability on the carbon economy of contrasting common bean (Phaseolus vulgaris L.) genotypes. J. Exp. Bot. 2001, 52, 329-339.
111. Williamson, L.C.; Ribrioux, S.P.C.P.; Fitter, A.H.; Leyser, H.M.O. Phosphate availability regulates root system architecture in Arabidopsis. Plant Physiol. 2001, 126, 875-882.
112. Lynch, J.P.; Brown, K.M. Topsoil foraging—An architectural adaptation of plants to low phosphorus availability. Plant Soil 2001, 237, 225-237.
113. Perez-Torres, C.; Lopez-Bucio, J.; Cruz-Ramirez, A.; Ibarra-Laclette, E.; Dharmasiri, S.; Estelle, M.; Herrera-Estrella, L. Phosphate availability alters lateral root development in Arabidopsis by modulating auxin sensitivity via a mechanism involving the TIR1 auxin receptor. Plant Cell 2008, 20, 3258-3272.
114. Liao, H.; Rubio, G.; Yan, X.L.; Cao, A.Q.; Brown, K.M.; Lynch, J.P. Effect of phosphorus availability on basal root shallowness in common bean. Plant Soil 2001, 232, 69-79.
115. Jungk, A. Root hairs and the acquisition of plant nutrients from soil. J. Plant Nutr. Soil Sci. 2001, 164, 121-129.
116. Narang, R.A.; Bruene, A.; Altmann, T. Analysis of phosphate acquisition efficiency in different Arabidopsis accessions. Plant Physiol. 2000, 124, 1786-1799.

117. Gahoonia, T.S.; Nielsen, N.E.; Joshi, P.A.; Jahoor, A. A root hairless barley mutant for elucidating genetic of root hairs and phosphorus uptake. Plant Soil 2001, 235, 211-219.
118. Gahoonia, T.S.; Nielsen, N.E. Barley genotypes with long root hairs sustain high grain yields in low-P field. Plant Soil 2004, 262, 55-62.
119. Gahoonia, T.S.; Nielsen, N.E. Direct evidence on participation of root hairs in phosphorus (32P) uptake from soil. Plant Soil 1998, 198, 147-152.
120. Yan, X.L.; Liao, H.; Beebe, S.E.; Blair, M.W.; Lynch, J.P. QTL mapping of root hair and acid exudation traits and their relationship to phosphorus uptake in common bean. Plant Soil 2004, 265, 17-29.
121. Grierson, C.S.; Parker, J.S.; Kemp, A.C. Arabidopsis genes with roles in root hair development. J. Plant Nutr. Soil Sci. 2001, 164, 131-140.
122. Zhu, J.M.; Kaeppler, S.M.; Lynch, J.P. Mapping of QTL controlling root hair length in maize (Zea mays L.) under phosphorus deficiency. Plant Soil 2005, 270, 299-310.
123. Vance, C.P.; Uhde-Stone, C.; Allan, D.L. Phosphorus acquisition and use: Critical adaptations by plants for securing a nonrenewable resource. New Phytol. 2003, 157, 423-447.
124. Ward, C.L.; Kleinert, A.; Scortecci, K.C.; Benedito, V.A.; Valentine, A.J. Phosphorus-deficiency reduces aluminium toxicity by altering uptake and metabolism of root zone carbon dioxide. J. Plant Physiol. 2011, 168, 459-465.
125. Mudge, S.R.; Rae, A.L.; Diatloff, E.; Smith, F.W. Expression analysis suggests novel roles for members of the Pht1 family of phosphate transporters in Arabidopsis. Plant J. 2002, 31, 341-353.
126. Ramaekers, L.; Remans, R.; Rao, I.M.; Blair, M.W.; Vanderleyden, J. Strategies for improving phosphorus acquisition efficiency of crop plants. Field Crops Res. 2010, 117, 169-176.
127. Shin, H.; Shin, H.S.; Dewbre, G.R.; Harrison, M.J. Phosphate transport in Arabidopsis: Pht1;1 and Pht1;4 play a major role in phosphate acquisition from both low- and high-phosphate environments. Plant J. 2004, 39, 629-642.
128. Xu, G.; Chague, V.; Melamed-Bessudo, C.; Kapulnik, Y.; Jain, A.; Raghothama, K.G.; Levy, A.A.; Silber, A. Functional characterization of LePT4: A phosphate transporter in tomato with mycorrhiza-enhanced expression. J. Exp. Bot. 2007, 58, 2491-2501.
129. Rengel, Z.; Marschner, P. Nutrient availability and management in the rhizosphere: Exploiting genotypic differences. New Phytol. 2005, 168, 305-312.
130. Hinsinger, P. Bioavailability of soil inorganic p in the rhizosphere as affected by root-induced chemical changes: A review. Plant Soil 2001, 237, 173-195.
131. Ryan, P.R.; Delhaize, E.; Jones, D.L. Function and mechanism of organic anion exudation from plant roots. Annu. Rev. Plant Physiol. Plant Mol. Biol. 2001, 52, 527-560.
132. Keerthisinghe, G.; Hocking, P.J.; Ryan, P.R.; Delhaize, E. Effect of phosphorus supply on the formation and function of proteoid roots of white lupin (Lupinus albus L.). Plant Cell Environ. 1998, 21, 467-478.
133. Neumann, G.; Romheld, V. Root excretion of carboxylic acids and protons in phosphorus-deficient plants. Plant Soil 1999, 211, 121-130.

134. Yan, F.; Zhu, Y.Y.; Muller, C.; Zorb, C.; Schubert, S. Adaptation of H+-pumping and plasma membrane H+ ATPase activity in proteoid roots of white lupin under phosphate deficiency. Plant Physiol. 2002, 129, 50-63.
135. Li, L.; Liu, C.; Lian, X. Gene expression profiles in rice roots under low phosphorus stress. Plant Mol. Biol. 2010, 72, 423-432.
136. Hayman, D.S.; Mosse, B. Plant Growth responses to vesicular-arbuscular mycorrhiza. I. Growth of endogone-inoculated plants in phosphate deficient soils. New Phytol. 1971, 70, 19-27.
137. Bucher, M. Functional biology of plant phosphate uptake at root and mycorrhiza interfaces. New Phytol. 2007, 173, 11-26.
138. Shenoy, V.V.; Kalagudi, G.M. Enhancing plant phosphorus use efficiency for sustainable cropping. Biotechnol. Adv. 2005, 23, 501-513.
139. Douds, D.D.; Janke, R.R.; Peters, S.E. Vam fungus spore populations and colonization of roots of maize and soybean under conventional and low-input sustainable agriculture. Agric. Ecosyst. Environ. 1993, 43, 325-335.
140. Ryan, M.H.; Chilvers, G.A.; Dumaresq, D.C. Colonization of wheat by VA-mycorrhizal fungi was found to be higher on a farm managed in an organic manner than on a conventional neighbor. Plant Soil 1994, 160, 33-40.
141. Kaeppler, S.M.; Parke, J.L.; Mueller, S.M.; Senior, L.; Stuber, C.; Tracy, W.F. Variation among maize inbred lines and detection of quantitative trait loci for growth at low phosphorus and responsiveness to arbuscular mycorrhizal fungi. Crop Sci. 2000, 40, 358-364.
142. Oehl, F.; Sieverding, E.; Mader, P.; Dubois, D.; Ineichen, K.; Boller, T.; Wiemken, A. Impact of long-term conventional and organic farming on the diversity of arbuscular mycorrhizal fungi. Oecologia 2004, 138, 574-583.
143. Xavier, L.J.C.; Germida, J.J. Response of spring wheat cultivars to Glomus clarum NT4 in a P-deficient soil containing arbuscular mycorrhizal fungi. Can. J. Soil Sci. 1998, 78, 481-484.
144. Wissuwa, M.; Ae, N. Genotypic variation for tolerance to phosphorus deficiency in rice and the potential for its exploitation in rice improvement. Plant Breed. 2001, 120, 43-48.
145. Marschner, P.; Solaiman, Z.; Rengel, Z. Rhizosphere properties of Poaceae genotypes under P-limiting conditions. Plant Soil 2006, 283, 11-24.
146. Plaxton, W.C.; Carswell, M.C. Metabolic Aspects of the Phosphate Starvation Response in Plants. In Plant Responses to Environmental Stress: From Phytohormones to Genome Reorganization; Lerner, H.R., Ed.; Marcel-Dekker: New York, NY, USA, 1999; p. 350.
147. Wissuwa, M.; Mazzola, M.; Picard, C. Novel approaches in plant breeding for rhizosphere-related traits. Plant Soil 2009, 321, 409-430.
148. Dawson, J.C.; Murphy, K.M.; Jones, S.S. Decentralized selection and participatory approaches in plant breeding for low-input systems. Euphytica 2008, 160, 143-154.
149. Desclaux, J.C. Participatory Plant Breeding Methods for Organic Cereals. In Proceedings of the COST SUSVAR/ECO-PB Workshop on Organic Plant Breeding Strategies and the Use of Molecular Markers; Lammerts van Bueren, E.T., Ostergard, H., Eds.; Driebergen, The Netherlands, 17–19 January 2005; p. 17.

150. Allard, R.W.; Hansche, P.E. Some parameters of population variability and their implications in plant breeding. Adv. Agron. 1964, 16, 281-325.
151. Danquah, E.Y.; Barrett, J.A. Grain yield in composite cross five of barley: Effects of natural selection. J. Agric. Sci. 2002, 138, 171-176.
152. Villa, T.C.C.; Maxted, N.; Scholten, M.; Ford-Lloyd, B. Defining and Identifying Crop Landraces. Plant Genet. Resour. 2005, 3, 373-384.
153. Banziger, M.; Cooper, M. Breeding for low input conditions and consequences for participatory plant breeding: Examples from tropical maize and wheat. Euphytica 2001, 122, 503-519.
154. Jalata, Z. GGE-biplot analysis of multi-environment yield trials of barley (Hordeium vulgare L.) genotypes in Southeastern Ethiopia Highlands. Int. J. Plant Breed. Genet. 2011, 5, 59-75.
155. Ceccarelli, S.; Grando, S.; Impiglia, A. Choice of selection strategy in breeding barley for stress environments. Euphytica 1998, 103, 307-318.
156. Atlin, G.N.; Frey, K.J. Selecting oat lines for yield in low-productivity environments. Crop Sci. 1990, 30, 556-561.
157. Abdelmula, A.A.; Link, W.; von Kittlitz, E.; Stelling, D. Heterosis and inheritance of drought tolerance in faba bean, Vicia faba L. Plant Breed. 1999, 118, 485-490.
158. Witcombe, J.R.; Joshi, A.; Goyal, S.N. Participatory plant breeding in maize: A case study from Gujarat, India. Euphytica 2003, 130, 413-422.
159. Sperling, L.; Ashby, J.; Weltzien, E.; Smith, M.; McGuire, S. Base-Broadening for Client-Oriented Impact: Insights from Participatory Plant Breeding Field Experience. In Broadening the Genetic Base of Crop Production; Cooper, H.D., Spillane, C., Hodgkin, T., Eds.; FAO-IPGRI: Rome, Italy, 2001; p. 419.
160. Heffner, E.L.; Sorrells, M.E.; Jannink, J. Genomic selection for crop improvement. Crop Sci. 2009, 49, 1-12.
161. Tester, M.; Langridge, P. Breeding technologies to increase crop production in a changing world. Science 2010, 327, 818-822.

CHAPTER 12

SPATIAL-TEMPORAL VARIATION OF POPULATION GROWTH AND SUSTAINABILITY OF FOOD GRAIN PRODUCTION IN WEST BENGAL, INDIA

SANJIT SARKAR AND KASTURI MONDAL

12.1 INTRODUCTION

Population increases in a geometrical ratio whereas food production increases only in arithmetic ratio. This fact of unbalanced growth pattern between population and food production became a priority discussion point in late eighteen century when Malthus predicted that population growth would outstrip food supply, causing great human suffering. When population continues rapidly within the limited resources, it has several adverse implications on earth, agriculture, bio-diversity, environment and population itself [1, 2]. Hence the most focused aim of government population policy is population stabilization, either by expanding family planning services or improving socio economic status of women [3]. Bucharest conference in 1974 emphasized to reduce the population growth by focusing

This chapter was originally published under the Creative Commons Attribution License. Sarkar S and Mondal K. Spatial–Temporal Variation of Population Growth and Sustainability of Food Grain Production in West Bengal, India. Journal of Settlements and Spatial Planning *3,1 (2012): pp. 35–42.*

both on fertility limitation and on related development aims. Immediate reduction in fertility do not guarantee the population stabilization, it is mostly depend on the age composition of the population. Population will be stabilized in a condition when numbers of women leaving the reproductive age group will equal to the numbers of women entering in the reproductive age group. Until this condition is achieved, population have tendency to increase which is called as 'population momentum'. Population dynamic simply means the short-terms and long-terms changes in the size and age structure of the population. It deals with the way population is affected by birth and death rates and by immigration and emigration. The linkage between population dynamic, food production and nutrition security is complex to generalize [4, 5]. There is no steady relationship between population growth and food production [6]. There have been large fluctuations in agricultural production. Though after mid 1980s, India achieved self-sufficiency in food grain production but there is no surety that grains in productivity would be sustained (Parikh, 2007). One concern is that Indian agriculture is mostly affected by climatic or natural barriers like drought, flood etc. And another concern is the environmental effects of High Yield Variety (HYV) technology. Use of chemical pesticides and fertilizers in agriculture may increase the yield capacity in short-term but it has adverse affect on soil fertile which may reduce the yield per capita in the future. Indian agriculture always remains as a subsistence type in nature but not as a commercial. Roy and Pal (2002) [8] showed that public investment in agriculture has declined considerably which might affect adversely in agriculture. The concern might be true. Many of the authors urged that agricultural growth rates in India are slowing down [9, 10, 11]. When growth rate of food grain production turns down, but population is hardly declining, then sustainability of food grain availability is a big question mark to us.

Though West Bengal is performing well in the food grain production but major concern is in the future sustainability of the food grain production. Because of the site advantage, being located on fertile Gangetic plain, West Bengal remains most densely populated state in India. According to 2011 population census of India, West Bengal population reached to 91.3 million with an additional increase of 11 million population since last census. With this large numbers of population it is very essential for the state to increase the current level of food grain production to ensure the quality and adequate diet for most humans. In West Bengal many people are malnourished today, especially children (38% under age five) and women. Though fertility of West Bengal reached to the replacement level (bellow 2.1) but the population momentum will guarantee a continued population growth for next few decades in the state. As population expands, the food problem will become increasingly sever, conceivably with numbers of malnourished. Considering of these facts, the paper will examine the temporal and spatial variation of population growth and food production in West Bengal.

12.2 DATA AND METHODOLOGY

The data for the present study has been gleaned from multiple sources in view of the array of different parameters to assess population dynamics, food production and nutrition security in West Bengal. The different sources are: Census of India; Statistical Abstract of West Bengal, 2008; and others published reports. Census data have been used to understand spatial and temporal pattern of population growth where annual exponential growth rate and decadal growth rate have been computed. Information regarding food grain production and land use pattern has been gathered form Statistical Abstract of West Bengal.

Percentage of cultivable land and percentage of net sown area were defined as the proportion of cultivable land and proportion of net sown area, respectively, to the total geographical area of particular administrative unit.

12.3 RESULTS AND DISCUSSION

12.3.1 POPULATION DYNAMICS IN WEST BENGAL

In the last century, population of West Bengal has increased more than 5 times than that of population in 1901. In 1901, the total population was 16.9 million which increased to 91.3 million in 2011, adding a total 74.4 million population in the last one hundred and ten years (table 1).

TABLE 1: Population dynamics and growth pattern in West Bengal, 1901–2011.

Year	Total population in India (million)	Population dynamics in West Bengal			
		Total population (million)	% share of national	Average annual exponential growth rate (percent)	Progressive growth rate over 1901
1901	238.4	16.9	7.1	-	-
1911	252.1	18.0	7.1	0.6	6.5
1921	251.3	17.5	6.9	-0.3	3.6
1931	279.0	18.9	6.8	0.8	11.8
1941	318.7	23.2	7.3	2.0	37.3
1951	361.1	26.3	7.3	1.3	55.6
1961	439.2	34.9	7.9	2.8	106.5
1971	548.2	44.3	8.1	2.4	162.1
1981	683.3	54.6	8.0	2.1	223.1
1991	846.4	68.0	8.0	2.2	302.4
2001	1028.7	80.2	7.8	1.7	374.6
2011*	1210.2	91.3	7.5	1.3	440.2

*Source: Census of India, * provisional 2011*

There was a fluctuating trend in the population growth in West Bengal. Till 1931, the annual exponential growth rates were negligible. It shows a negative growth rate during 1911 to 1921. During this time population declined marginally due to great influenza epidemic and two successive bad harvests.

The late eighteen and nineteen centuries saw the worst famines and these were bad enough to have a remarkable impact on the long term population growth of the country, especially in the half century between 1871–1921. As a part of the nation, West Bengal was also influenced by these famines. The Bengal famine of 1770 is estimated to have taken the live of nearly onethird of the population of the region. After 1921, there was continues growth trend in the population. West Bengal's population increased to 26.3 million in 1951 from 17.5 million in 1921. During this period the growth rates were low (average annual exponential growth rate was 1.3 percent) because of the high fertility and high mortality. Hence, natural increase in population was quite low, added only nine million population in these thirty years. During 1951-1981, the growth rate was very high, exceeding 2.4 percent average annual exponential growth rate. In this period the population increased to 54.6 million in 1981 from 26.3 million in 1951, adding a total of 28.3 million population. After independence mortality has declined very faster rate than fertility which broaden the gap between births and deaths and ultimately it contributes on the natural increase of the population. Besides the natural increase, effect of net migration was significant in the rapid growth of population in West Bengal. During 1981-2011, the growth rate showed a declined trend (average annual exponential growth rate 1.7 percent) over the previous period. Though there was a slight fluctuation between 1981 and 1991 but afterwards it showed a steady decline trend. In the last decade (2001-2011) the average annual exponential growth rate reached to 1.3 percent.

Figure 1 shows that along with the mortality, fertility also started to decline over the time and hence natural growth rate turned down remarkably. During 1981-83 natural growth rate was 21.9 per thousand of live births and it came down to 12.1 per thousand of live birth during 2005-2007.

Figure 2 shows that decadal growth rate of West Bengal declined since form 1971-1981. During 2001-2011, decadal growth rate declined to 13.9 percent which is much lower than the national growth rate (17.6 percent). Though the growth rate of population in West Bengal is lower than the national average but it is significantly higher than some of the South Indian states.

The growth rate of West Bengal started to decline since 1981-1991 whereas growth rate of Kerala began to decline nearly 30 years before than West Bengal since 1961-1971.

Now the major concern of population dynamics is that though fertility reached to the replacement level of fertility in West Bengal (TFR below 2.1) but it will not guarantee the population stabilization in the near future because of the impact of population momentum. This momentum in population will continue for some more years because high TFR in the past have resulted in a large proportion of population being currently in their reproductive age group.

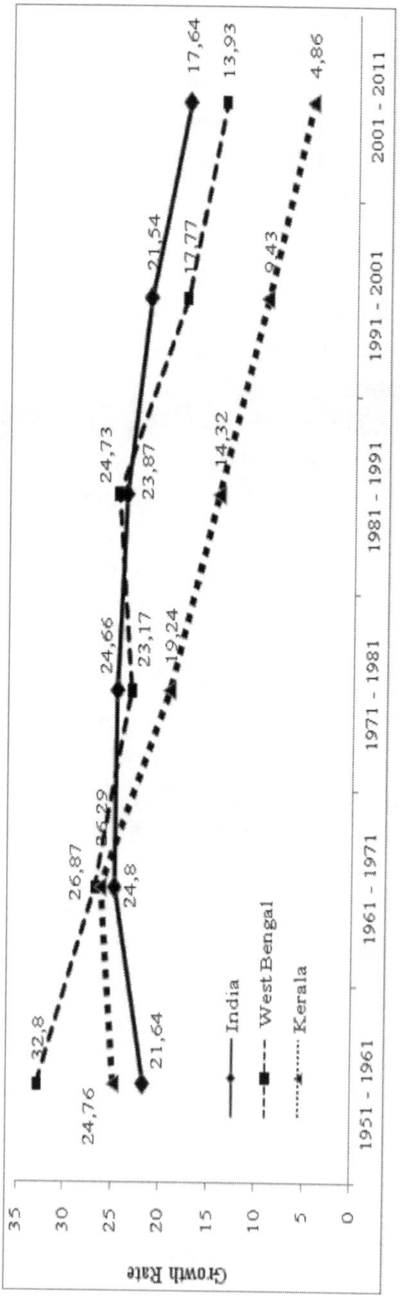

FIGURE 1: Trends in Natural Increase in West Bengal 1981-2009.

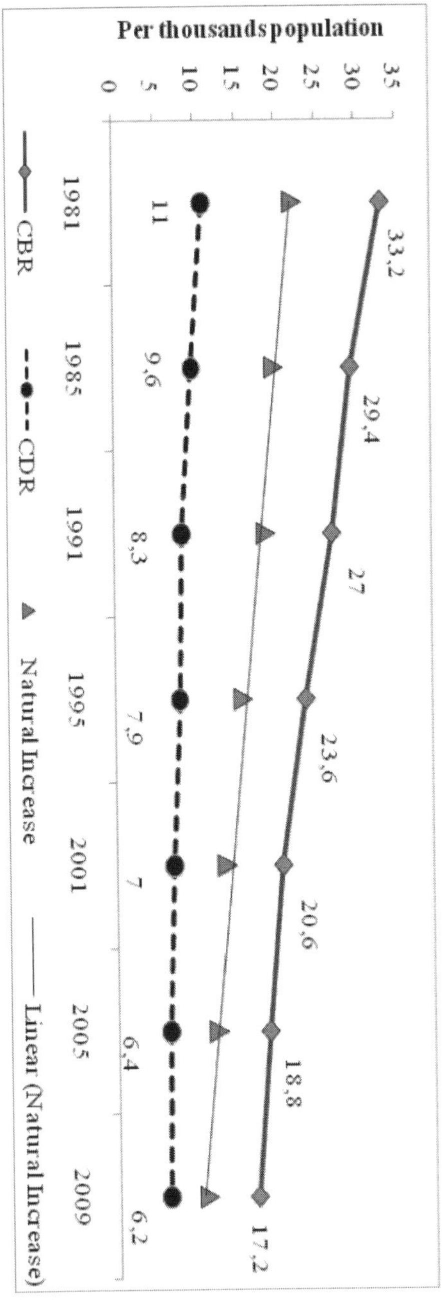

FIGURE 2: Trends of decadal growth rate in West Bengal, Kerala and India.

TABLE 2: Temporal and district wise growth pattern of population in West Bengal 1981-2011.

	Decadal absolute growth of population ('00000)			Change in absolute growth during 1981 -2011		Percent contribution to growth		
District Name	1981 - 1991	1991 - 2001	2001 - 2011	Absolute Change ('00000)	% Change	1981 - 1991	1991 - 2001	2001 - 2011
West Bengal	135.0	121.4	111.3	-23.7	-17.6	100	100	100
Burdwan	12.2	8.7	8.0	-4.1	-33.8	9.0	7.2	7.2
Birbhum	4.6	4.6	4.9	0.3	6.5	3.4	3.8	4.4
Bankura	4.3	3.9	4.0	-0.3	-6.0	3.2	3.2	3.6
Midnapur	15.9	13.1	14.0	-1.9	-12.0	11.8	10.8	12.6
Howrah	7.6	5.4	5.7	-2.0	-25.6	5.7	4.5	5.1
Hooghly	8.0	6.8	4.8	-3.2	-39.8	5.9	5.6	4.3
24-Pargonas (N)	-	16.5	11.5	-	-	-	13.6	10.4
24-Pargonas (S)	-	11.9	12.4	-	-	-	9.8	11.2
Kolkata	10.9	1.8	-0.9	-11.9	-108.6	8.1	1.5	-0.8
Nadia	8.9	7.5	5.6	-3.2	-36.4	6.6	6.2	5.1
Murshidabad	10.4	11.2	12.4	2.0	18.8	7.7	9.3	11.1
Dinazpur (N)	-	5.4	5.6	-	-	-	4.5	5.0
Dinazpur (S)	-	2.7	1.7	-	-	-	2.2	1.5
Malda	6.1	6.5	7.1	1.0	17.0	4.5	5.4	6.4
Jalpaiguri	5.9	6.0	4.7	-1.2	-20.4	4.3	5.0	4.2
Darjeeling	2.8	3.1	2.4	-0.4	-14.3	2.0	2.5	2.1
Coochbehar	4.0	3.1	3.4	-0.6	-13.8	3.0	2.5	3.1
Purulia	3.7	3.1	3.9	0.2	5.9	2.7	2.6	3.5

Source: Census of India; Note: 2011 data is the provisional census data.

12.3.2 SPATIAL GROWTH OF POPULATION IN WEST BENGAL

Spatial pattern of decadal absolute growth of population during 1981-2011 has been presented in table 2. During 1981-91, decadal absolute growth of West Bengal was recorded by 13.5 million population. Highest decadal growth rate was found for Midnapur district where nearly 1.6 million population increased in this decade and followed by Burdhaman (1.2 million), Kolkata (1.1 million) and Murshidabad (1.0 million) districts. These four districts collectively contribute almost 36 percent in the decadal absolute growth of the state during 1981-1991. In the next decade (1991-2001) the decadal absolute growth of population was 12.1 million which showed a reduction of 1.4 million population since previous decade. During this time the highest decadal absolute growth was fond for North-24 Pargonas (1.6 million population) followed by Midnapur (1.3 million), South 24 Pargonas (1.2 million) and Musshidabad (1.1 million) districts. These four districts together contribute 43 percent in the absolute decadal growth of the state during 1991-01. Most amazingly, Kolkata's population showed a stagnant absolute growth during 1991-2001, showing an additional increase of only 1.8 lack population. During 2001-11, West Bengal was recorded with the absolute growth of 11.1 million population which showed a farther reduction of one million population over the last decade. In this period, highest absolute growth found for Midnapur distrct (1.4 million population), followed by South 24 Pargonas (1.2 million), Murshidabad (1.2 million) and North 24 Pargonas (1.1 million). These four districts together contribute 45 percent in the decadal absolute growth of the state.

Figure 3 shows the districts wise variation of decadal growth rates during 1991-01 and 2001-11 respectively. Three types of area may be demarcated: Districts with high growth rate (more than 20 percent); Districts with moderate growth rate (10-20 percent); District with slow growth rate (less than 10 percent). During 1991-2001, total eight district out of nineteen showed higher decadal growth rates, among them Uttar Dinazpur (28.7 percent), Maldha (24.8 percent) and Murshidabad (23.7 percent) are significant. These districts are mainly located on the border of Bangladesh, Nepal and Bihar. Hence, regular in-migration from these surrounding states and countries contributed a lot to reach the higher population growth rate of these districts. Nine districts showed a moderate growth rates, whereas Kolkata showed a slow decadal growth rate with only four percent growth rate. The growth rate of population declined significantly during 2001-11 over the last decade. Decadal growth rate of the state reached to 13.9 percent from 17.8 percent, a decline of four percent, during last two decadal periods. In this period (2001-11) most of the districts show moderate decadal growth rate (10-20 percent), whereas Kolkata experiences a negative (-2 percent) decadal growth pattern.

The significant reduction in the population growth rate in recent decade is mainly due to the significant decline in the natural increase of population and strict control over the immigrations. Illegal migration from neighbouring country has been a major problem for Bengal since independence. After 1971 large numbers of Bangladeshi nationals have crossed over to Bengal in search of livelihood. Bangladeshi migrants have significant influence in the population growth of West Bengal. The rate of population growth in the nine Bengal districts that share their borders with Bangladesh has come down in the last decade.

The reduced growth rate of population in the districts such as Jalpaiguri, North and South Dinazpur, Nadia and North 24 Pargonas is the high light of this recent census. This clearly indicates that illegal migration has been checked substantially by fencing more border area and steeping up vigil on illegal migration. Besides the administrative cheek, demographic change and socio economic development of Bangladesh are also significant to reduce the influx form Bangladesh. Demographers said that reduction of natural increase of population in Bangladesh was also a reason for the reduction of in-migration.

12.3.3 FOOD GRAIN PRODUCTION IN WEST BENGAL

West Bengal is situated in the most fertile land of lower Gangetic Plain and hence agriculture plays a pivotal role in the state's economy and nearly three out of every four persons is directly or indirectly involved in agriculture. In spite of an agriculture-dependent state, West Bengal was still dependent on the central government for meeting the domestic food demands till 1980s. However, there has been a significant spurt in the food grain production and now the state has a surplus of food grain. The total food production in the State in 2007- 2008 was 16,060 thousand tonnes (table 3).

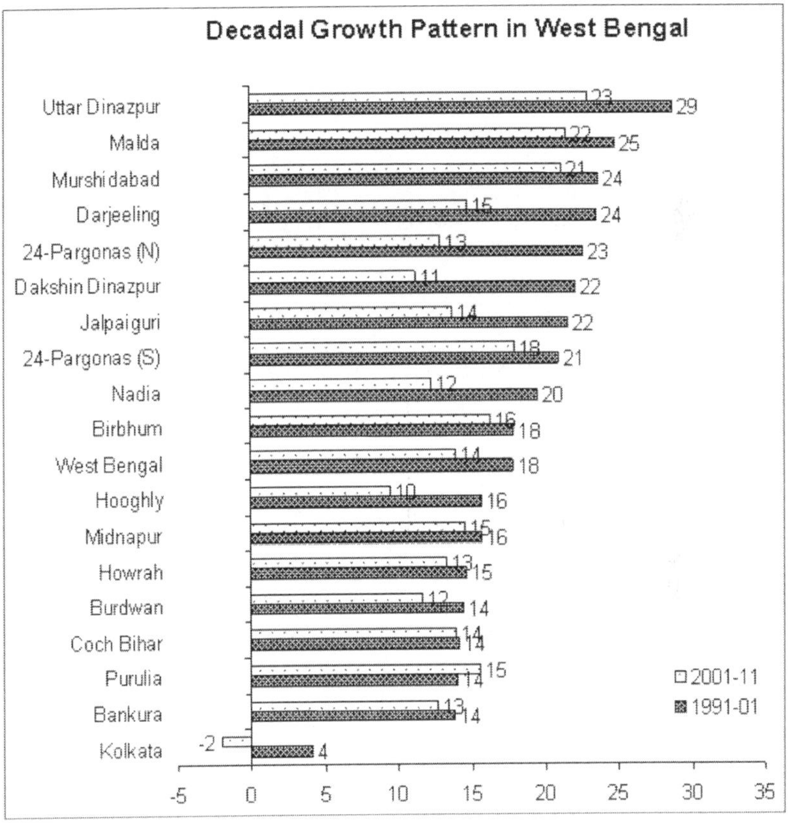

FIGURE 3: District wise decadal growth rate in West Bengal, 1991-01 & 2001-11.

TABLE 3: Temporal and district level variation in food grain production in West Bengal, 1990-2008.

Districts	Total food grain (in thousand tonnes)			Temporal change (in thousand tonnes)			Percent contribution to food grain production		
	1990-01	2000-01	2007-08	1990-01 2000-01	2000-01 2007-08	1990-01 2007-08	1990-01	2000-01	2007-08
Burdwan	1,427.5	1,592.9	1,866.3	165.4	273.4	438.8	12.7	11.5	11.6
Birbhum	844.1	892.1	1,334.4	48	442.3	490.3	7.5	6.5	8.3
Bankura	866.6	1,016.7	1,182.8	150.1	166.1	316.2	7.7	7.4	7.4
Midnapur	1,558.7	2,638.7	2,652.5	1080	13.8	1093.8	13.8	19.1	16.5
Howrah	234.9	226.7	260.8	-8.2	34.1	25.9	2.1	1.6	1.6
Hooghly	612.7	508.4	846.8	-104.3	338.4	234.1	5.4	3.7	5.3
24-Pargonas (N)	674.2	685.2	769.2	11	84	95	6.0	5.0	4.8
24-Pargonas (S)	513.9	879.5	812.8	365.6	-66.7	298.9	4.6	6.4	5.1
Nadia	830.1	843.5	829.8	13.4	-13.7	-0.3	7.4	6.1	5.2
Murshidabad	995	979.8	1,518.2	-15.2	538.4	523.2	8.8	7.1	9.5
Dinazpur (N)	846.8*	721.1	846	369.3*	124.9	522.9*	7.5*	5.2	5.3
Dinazpur (S)	495	523.7	-	28.7	-	-	3.6	3.3	
Malda	633.2	688.7	667.3	55.5	-21.4	34.1	5.6	5.0	4.2
Jalpaiguri	275.2	441.5	439.4	166.3	-2.1	164.2	2.4	3.2	2.7
Darjeeling	121.1	114.6	116.5	-6.5	1.9	-4.6	1.1	0.8	0.7
Coochbehar	432.4	574.6	610	142.2	35.4	177.6	3.8	4.2	3.8
Purulia	403.7	516.2	784	112.5	267.8	380.3	3.6	3.7	4.9
West Bengal	11,270	13,815	16,060.5	2,545.1	2,245.3	4,790.4	100	100.0	100.0

*Source: Statistical Abstract of West Bengal, 2008; Note: * = Combined Dinazpur*

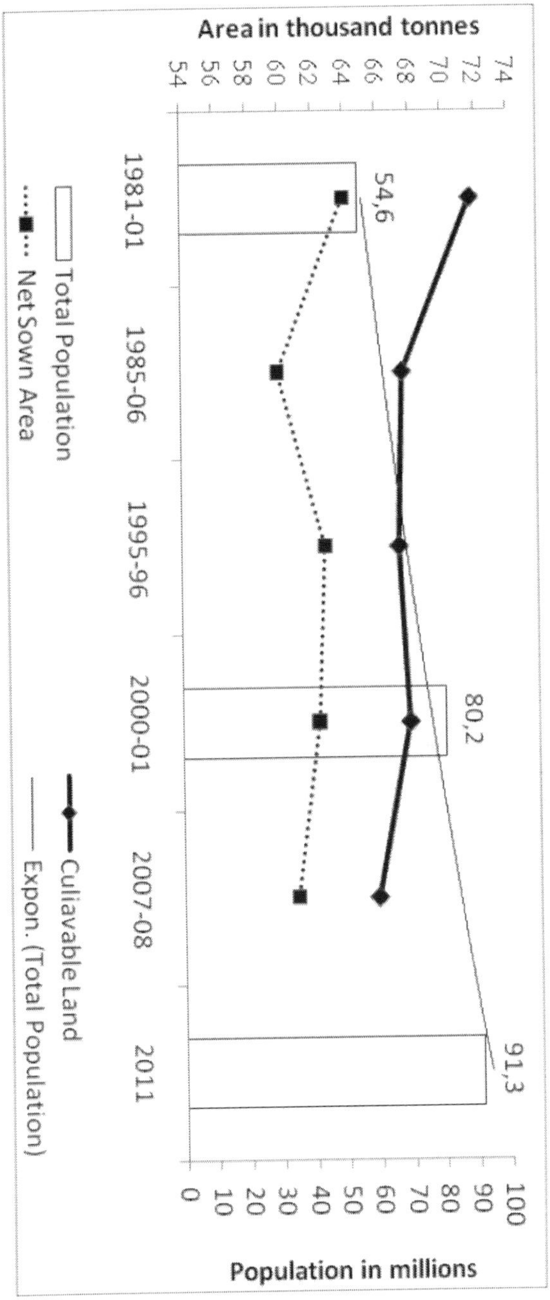

FIGURE 4: Trend of Food production and Food Requirement Gap in West Bengal.

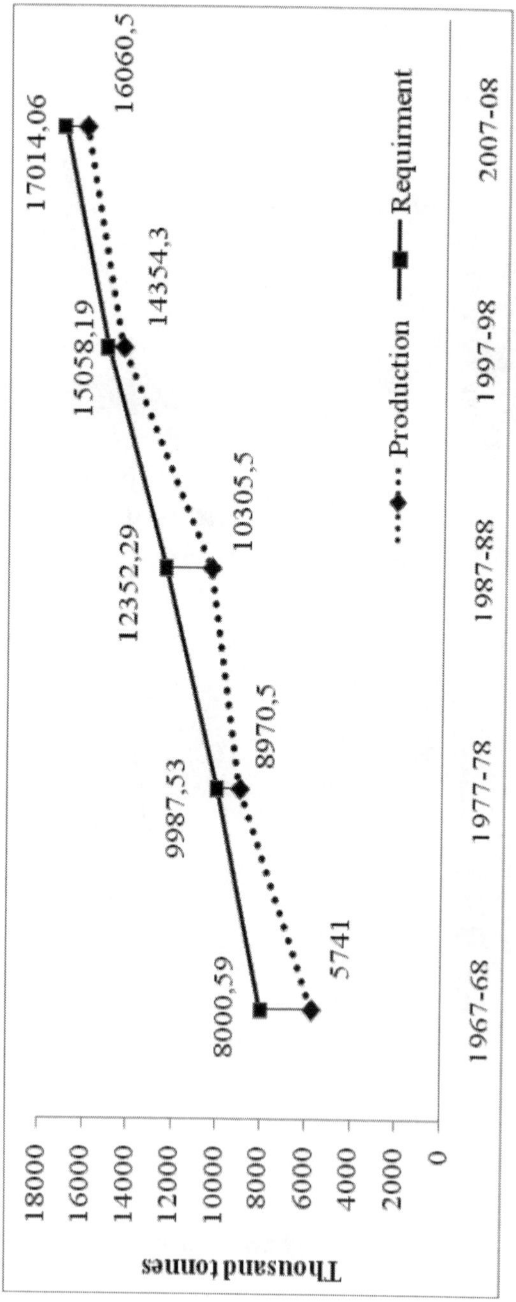

FIGURE 5: Relationship between population growth pattern and availability of agricultural land in West Bengal.

During 2007-08, the production of rice was 14719.2 thousand tonnes, of wheat 917.3 thousand tonnes and of pulses 158.0 thousand tonnes respectively. There was substantial increase in the food grain production in West Bengal in the last two decades. Table 3 shows a spatial-temporal variation of food grain production in West Bengal. In 1990-91 the state's food grain production was 11270 thousand tonnes which reached to 16060 thousand tonnes in 2007-08. During 1990-91 to 2007-08 the additional increment in food grain production was 4790 thousand tonnes.

Temporal change in food grain production varies significantly over the districts. Among all the districts, Midnapur shows the highest increase in the food grain production during 1990-91 to 2007-08, adding a total 1093 thousand tonnes of food grain during the period and followed by Murshidabad (523 thousand tonnes), Dinazpur (522 thousand tonnes) and Bankura district (490 thousand tonnes). There is spatial inequality in the food grain production in West Bengal. More than fifty percent of the state production is confined only in the five districts out of nineteen. During 1990-91 the top five food grain producing districts were Midnapur (13.8 percent), Burdwan (12.7 percent), Murshidabad (8.8 percent), Bankura (7.7 percent) and Birbhum (7.5 percent). These five districts together contributed 50 percent of the state production during 1990-91 whereas during 2007-08 the contribution increased to 53 percent. Food grain production in West Bengal increased significantly over the time.

The gap between food grain production and requirement decreased significantly in the recent decades (fig 4) but the current level of food grain production is not sufficient enough to meet the domestic food requirement. Fig 4 shows that during 2007-2008, total food grain production was 16060 thousand tonnes against its requirement of 17014 thousand tonnes.

12.3.4 POPULATION GROWTH AND AVAILABILITY OF LAND FOR AGRICULTURE

Rapid growth of population in a geographical area affect adversely on the land, especially on the cultivable land. It creates pressure on the cultivable land by reducing its area. Production can decline marginally when the land crosses its caring capacity due to over population.

TABLE 4: District wise variation of percentage of cultivable land and net sown area in West Bengal, 1995-96 to 2007-08.

	1995-96		2007-08		Percent points change during 1995-96 to 2007-08	
State & District	% cultivable land	% net sown area	% cultivable land	% net sown area	% cultivable land	% net sown area
West Bengal	67.4	62.8	65.9	61.0	-1.4	-1.8
Burdwan	69.7	66.7	67.3	64.7	-2.4	-2.0
Birbhum	75.0	68.9	75.0	70.6	0.0	1.7
Bankura	59.6	53.1	56.6	50.2	-3.0	-2.9
Midnapur	66.5	63.8	67.5	64.3	1.0	0.4
Howrah	72.0	62.9	63.1	58.2	-8.9	-4.7
Hooghly	74.9	73.0	71.3	70.2	-3.5	-2.8
24-Pargonas (N)	71.4	67.7	68.5	67.1	-2.9	-0.7
24-Pargonas (S)	42.3	41.2	40.4	39.2	-1.8	-2.0
Nadia	81.4	76.2	77.0	74.0	-4.4	-2.2
Murshidabad	79.6	76.5	75.5	74.9	-4.1	-1.6
Dinazpur*	88.1	85.3	87.8	86.4	-0.4	1.1
Malda	82.4	76.4	75.8	56.7	-6.6	-19.7
Jalpaiguri	53.8	51.6	57.1	53.7	3.3	2.1
Darjeeling	50.6	44.8	49.6	43.2	-1.0	-1.6
Coochbehar	79.7	75.8	77.9	74.8	-1.8	-1.0
Purulia	71.0	54.2	71.3	50.0	0.3	-4.2

Note: Cultivable land includes the following categories: Permanent pastures and grazing land; Misc.tree crops and groves; Culturable waste; Fellow land; Current fellow; Net sown area.

Availability of cultivable land and net sown area has declined remarkably in West Bengal due to increase in population (fig. 5). Table 4 shows that proportion of cultivable land has declined from 67.4 percent to 66 percent during 1995-96 to 2007-08 in West Bengal. Proportion of net sown area also declined from 63 percent to 61 percent during the same period. Highest decline in cultivable land was recorded in Howrah district (8.9 percent points) during 1995-96 to 2007-08, followed by Malda (6.6 percent points) and Nadia (4.4 percent points). A significant decline in the net sown area was recorded in Malda district from 76 percent to 57 percent, accounting a sharp decline of 19 percent points during 1995-96 to 2007-08. There are substantial difference between availability of cultivable land and land used for cultivation. Nearly 66 percent of total geographical area is available for cultivation but only 61 percent land is used for cultivation purpose. The gap is highest in Purulia district where 71 percent of total geographical area is cultivable land but only 50 percent is used for cultivation. Land utilization of cultivation is very poor in South 24 Pargonas (39 percent), Darjeeling (43 percent), Purulia and Bankura (50 percent).

12.4 CONCLUSION

It may conclude from the foregoing discussion that absolute numbers of population in West Bengal increased over the last two decades but growth rate has declined significantly. Average annual growth rate was 2.2 percent in 1981-91 which reduced to 1.3 percent during 2001-11. A significant reduction in the growth rate is also recorded in some of the districts in West Bengal. During 1991-01, total six districts showed a higher annual exponential growth rate (more than 2.0 percent) but during 2001-11 only one district showed higher annual exponential growth rate. Surprisingly, Kolkata showed a negative growth rate during 2001-11. Sudden decline in the population of Kolkata may be a result of low natural increase of population or of huge out migration from the city or may be a result of both. Though growth rate of population is decreasing but absolute growth in the population is one of the major concerns of population dynamics in West Bengal. It increases the population density, reduces the availability of cultivable land and increases the domestic demand for food. The population growth in West Bengal has significant association with food grain production and agriculture. The cropping intensity has gone up remarkably due to the population growth and shrinking in the cultivable land. The food grain production has increased significantly in West Bengal but it is not sufficient enough to meet the domestic food requirement. So, there is urgent need to improve the current level of food grain production. On the other hand, food grain production in West Bengal is showing a slow growth pattern since last few years. Hence, food security and sustainability of food grain production will be a major concern in the very near future in West Bengal.

REFERENCES

1. Kodekodi, K. G. (2007), On Linkage between Population and Environment: Some Evidences from India, In: Prakasam, C.P., and Bhagat, R.B. [editors] Population and Environmental Linkage, Rawat Publication, New Delhi, India, pp.3-19.
2. Dasgupta, P. (2003), Population, Poverty and Natural Environment, In: Maler, K.G., and Vincent, J. R. [editors]. Handbook of Environmental Econimics, Edward Elgar, Cheltenhum, pp. 191-247.
3. *** (2000), National Population Policy 2000, Planning Commission, Government of India, India.
4. Sawant, S. D., Achuthan C. V. (1995), Agricultural Growth Across Crops and Regions: Emerging Trends and Patterns, In: Economic and Political Weekly, March 25.
5. Southgate, D. (2009), Population Growth, Increase in Agriculture Production and Food Price, Electronic Journal of Sustainable Development, Vol 1(3).
6. Swaminathan, M. (2010), Population and Food Security, In: Shiva Kumar,A., Panda,P., and Rajani R. V. (editors). Handbook of Population and Development in India, Oxford University Press, New Delhi, India, pp- 50-56.
7. Parikh, A. (2007), Population vs Food: Who is winning the race, , In: Prakasam, C.P., and Bhagat, R.B. [editors] Population and Environmental Linkage, Rawat Publication, New Delhi, India, pp.77-87.
8. Roy, B. C., Pal, S. (2002), Investment, Agricultural Productivity and Rural Poverty in India: A State Level Analysis, Indian Journal of Agricultural Economics, Vol. (57(4), pp. 653-678.
9. Rao, V. M., Jeromi, P. D. (2000), Modernising Indian Agriculture: Priority Tasks and Critical Policies, Development Research Group, Study No 21, Department of Economic Analysis and Policy, Reserve Bank of India.
10. Dev, M. (1998), Regional Variation in Agricultural Performance in the Last Two Decades, Indian Journal of Agricultural Economics, Vol.53: No 1.
11. Mallik, J. K. (1997), Growth of Agriculture in Independent India: 50 Years and After, RBI occupational papers, Vol. 18, No. 2 and 3 special issue.

AUTHOR NOTES

CHAPTER 1

Acknowledgments

The research was supported by Beijing Natural Science Foundation (8122020). The authors are grateful to the Beijing Municipal Bureau of Water Management for their water sampling and analysis, and we would like to thank all reviewers and editors for their valuable comments and suggestions during the review process.

Conflict of Interest

The authors declare no conflict of interest.

CHAPTER 2

Acknowledgments

Co-funding of the project leading to these results by the European Commission within the 7th Framework Programme under Grant Number 213154 is kindly acknowledged.

CHAPTER 3

Acknowledgments

The authors acknowledge funding from the Ontario Ministry of Agriculture, Food and Rural Affairs, and assistance of Mike Zink, Paul Voroney, Ivan O'Halloran, Jessica Turnbull, summer students, and Jonathan Gorham.

CHAPTER 5

Funding

This study was financially supported by the German Federal Ministry for Education and Research, Projektträger DLR (projects MEX 06/003 and 01DN12067) and CONACYT (Concejo Nacional de Ciencia Y Tecnologia) J110.394 and the BMBF project MÄQNU 03MS642H. The funders had no role in study design, data collection and analysis, decision to publish, or preparation of the manuscript.

Competing Interests

The authors have declared that no competing interests exist.

Acknowledgments

We thank Professor Siegfried Kropf at the Institute for Biometry and Medical informatics, Otto von Guericke University for his help on statistical analyses and Ilse-Marie Jungkurth for proofreading the manuscript very carefully.

Author Contributions

Constructive discussion and sharp criticism to improve the manuscript: HH. Conceived and designed the experiments: KS CCT TC GLA. Performed the experiments: YMP NW ABD AC. Analyzed the data: G-CD HH. Contributed reagents/materials/analysis tools: YMP NW CCT TC G-CD. Wrote the paper: G-CD KS CCT TC GLA.

CHAPTER 6

Conflict of Interest

The authors declare no conflict of interest.

CHAPTER 7

Acknowledgments

The authors gratefully acknowledge the financial support from the National Natural Science Foundation of China (No. 70873013 & No. 71273039).

CHAPTER 11

Conflict of Interest

The authors declare no conflict of interest.

Acknowledgments

Scientific article No. 3117 of the West Virginia Agricultural and Forestry Experiment Station, Morgantown.

INDEX

A

B

C

D

P

T

U

V

W

Y